T0214650

Fundamentals of Power Systems Analysis 1

A supplementary book on power systems and their points is necessary for every successful student because the main books contain so much information. The supplementary book should include a summary, many tests, and an explanation of the answers.

The structure in *Fundamentals of Power Systems Analysis 1: Problems and Solutions* is very helpful for re-reading and summarizing the information. This book can help you increase your study speed and master the important lessons if you are in the last few months of the semester and have not studied.

- This book is styled after national exams, with many varied tests with complete descriptive answers
- This book covers everything you need to know about power systems analysis
- A comprehensive and detailed examination of each image and figure has been conducted in this book
- Students will be able to review points more quickly. It is particularly helpful before exams or national tests when you are under stress. It has the main advantage of providing an analysis of concepts and their combination. This allows students to better answer questions derived from several other subjects in a combined manner.

Fundamentals of Power Systems Analysis 1
Problems and Solutions

Mostafa Eidiani and Vahideh Heidari

CRC Press
Taylor & Francis Group
Boca Raton London New York

CRC Press is an imprint of the
Taylor & Francis Group, an **informa** business

First edition published 2024
by CRC Press
2385 Executive Center Drive, Suite 320, Boca Raton, FL 33431

and by CRC Press
4 Park Square, Milton Park, Abingdon, Oxon, OX14 4RN

CRC Press is an imprint of Taylor & Francis Group, LLC

ISBN: 9781032495637 (hbk)
ISBN: 9781032495613 (pbk)
ISBN: 9781003394433 (ebk)
ISBN: 9781032554624 (ebk+)

DOI: 10.1201/9781003394433

Typeset in Palatino
by codeMantra

Eidiani dedicated this book to his beloveds, Elham and Hosna.

Mrs. Heidari presented this book to her husband.

Contents

Preface

This enhanced version includes an additional feature to provide a greater understanding of the text. It includes over 300 yes/no quizzes to test the comprehension of the readers.

This book is recommended for electrical (power, telecommunications, control, and electronics) and computer engineering students as supplementary material for reviewing power systems 1. The formulas and concepts in this book can be accessed quickly as a handbook or fastbook.

A supplementary book on power systems and their points is necessary for every successful student because the main books contain so much information. The supplementary book should have a summary, many tests, and an explanation of the answers. There is a descriptive answer sheet in this book that is very helpful for re-reading and summarizing the information. This book can help you increase your study speed and master the most crucial lessons if you are in the last few months of the semester.

This book has the following advantages:

- In the style of national exams, there are many and varied tests with fully descriptive answers.
- This book covers everything you need to know about power systems analysis.
- For a faster review of points, especially during times of stress such as before examinations or national tests.
- Analyzing the concepts and combining them to better solve questions that are derived from several other courses in an integrated manner.
- A detailed and comprehensive examination of each book's image and figure points.

This book provides an overview of power systems that is both complete and convenient for students, engineers, and teachers. Students can use it as supplementary material to review the course in the power systems analysis course for undergraduates.

Finally, I want to express my gratitude to all of my professors: Prof. Mohammad Hassan Modir Shanehchi, Prof. Ebrahim Vaahedi, Prof. Mehdi Ehsan, Prof. Hossein Javidi, Prof. Reza Ghazi, Prof. Mehrdad Abedi, Prof. Mostafa Parniani, Prof. Heidarali Shayanfar, Prof. Ali Abbaspour, Prof. Ali Piravi, and Prof. Mahmud Fotuhi Firuzabad.

Authors

Mostafa Eidiani (StM'98-SM'16) was born in Mashhad, Iran. He received his B.Eng. (with distinction) and M.Eng degree in Electrical Engineering from Ferdowsi University of Mashhad, Iran in 1995, 1997, respectively and his Ph.D. from the Science and Research Branch, Islamic Azad University, Tehran, Iran in 2004. He is a senior member of IEEE in 2016 and director of the Iranian Association of Electrical and Electronics Engineers (Khorasan Branch) in 2017. He was promoted from assistant professor to associate professor in 2016. His research interests include renewable energy integration, power system control, transient and voltage stability, power system simulation, and DIgSILENT Powerfactory.

He has authored or co-authored 7 technical books, 6 chapter books, 40 journal papers, and 110 technical conference proceedings. In addition, he has conducted more than 30 research projects with Iranian power companies.

Dr. Eidiani is an editor-in-chief for *International Journal of Energy Security and Sustainable Energy* (www.ijesse.ir), Editorial Board for IET *Journal of Engineering, International Journal of Applied Power Engineering, Journal on Advanced Research in Electrical Engineering, Majlesi Journal of Energy Management, Electrical Asre Magazine, Recent Advances in Engineering & Technology, Sustainable Engineering and Innovation, and Energy and Power Engineering.* He has been a board member of Khorasan Electric Generation Company.

Vahideh Heidari was born in Mashhad, Iran in 1990. She received her B.Eng degree in Electrical Engineering of Telecommunication from Imam Reza International University of Mashhad, Iran in 2012 and her M.Eng. degree (with distinction) in Electrical Engineering of Control from Khorasan Institute of Higher Education of Mashhad, Iran in 2016. During the master's period, three articles from this student have been accepted, and one of these articles was published in the Control Journal of Khajeh Nasir University of Technology of Tehran, Iran in 2016.

1

Introduction to the Power System

Part One: Lesson Summary

1.1 Introduction

This chapter introduces the reader to the fundamental ideas of power system analysis. Sinusoidal steady-state, phasor representation, power definition, concepts of leading and lagging phase, concepts of production and consumption power, types of load models, power and impedance triangles, Y-Δ or star-delta transform, three-phase systems, power factor correction, per-unit, transmission power, generator and motor connected, and relevant numerical approximations are some of the concepts discussed in this chapter. Three-phase systems are believed to be balanced in the study of power systems.

Single-phase analysis may be used to examine these balanced three-phase systems. By combining an inductance and a capacitor, large single-phase loads that generate severe phase imbalances may be transformed into three balanced phases at once.

Voltage and frequency fluctuations affect the quantity of electric energy consumed. The effects of voltage and frequency on motor and impedance loads are depicted.

Power transmission formulae are also completely described in this section. The link between active and reactive power, production and consumption methods, and their varied modes has been studied.

1.2 Sinusoidal Steady-State

The instantaneous voltage and current:

$$i(t) = I_m \cos(\omega t + \theta_i) \quad v(t) = V_m \cos(\omega t + \theta_v) \tag{1.1}$$

DOI: 10.1201/9781003394433-1

The RMS value or effective value of a voltage or current:

$$V_e = V = \frac{V_m}{\sqrt{2}}, \quad I_e = I = \frac{I_m}{\sqrt{2}} \tag{1.2}$$

1.3 Phasor Representation

It must be determined from the question form which type of display was used. Electrical circuit type:

$$V = |V| \angle \theta_v, \quad I = |I| \angle \theta_i \tag{1.3}$$

Power system type:

$$\hat{V} = V \angle \theta_v, \quad \hat{I} = I \angle \theta_i \tag{1.4}$$

1.4 Power Definition

The instantaneous power:

$$p(t) = v(t) \cdot i(t), \, p(t) = V \cdot I \cdot \cos\varphi \left(1 + \cos(2\omega t)\right) + V \cdot I \cdot \sin(\varphi)\sin(2\omega t) \tag{1.5}$$

The angle between voltage and current:

$$\varphi = \theta_v - \theta_i \tag{1.6}$$

The average power:

$$P_{\text{av}} = \frac{1}{T} \int_0^T p(t) \cdot dt \tag{1.7}$$

The real or active power (Watt or W):

$$P = P_{\text{av}} = V \cdot I \cdot \cos(\varphi) \tag{1.8}$$

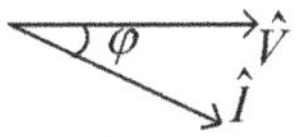

FIGURE 1.1
Phasor diagram of lagging load.

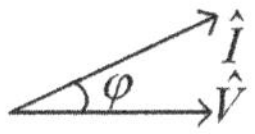

FIGURE 1.2
Phasor diagram of leading load.

The imaginary or reactive power (Volt-Ampere reactive or VAr):

$$Q = V \cdot I \sin(\varphi) \tag{1.9}$$

The complex power (Volt-Ampere or VA):

$$S = P + jQ = \hat{V} \cdot \hat{I}^* = V \cdot I \angle \varphi = |S| \angle \varphi = \left(\sqrt{P^2 + Q^2}\right) \angle \varphi \tag{1.10}$$

The apparent power or the magnitude of the complex power (VA):

$$|S| = \sqrt{P^2 + Q^2}$$

1.5 Concepts of Leading and Lagging Phase

Assume voltage as the reference.

An *inductive* load has a current angle *less* than its voltage angle, so the current *lags* the voltage and the power factor is thus *lagging* (Figure 1.1).

A *capacitive* load has a current angle *much larger* than its voltage angle, so the current *leads* the voltage and the power factor is thus *leading* (Figure 1.2).

1.6 Concepts of Production and Consumption Power

The following power relationship exists in both producer and consumer cases.

$$S = P + jQ = \hat{V} \cdot \hat{I}^* = V \cdot I \angle \varphi$$

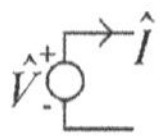

FIGURE 1.3
Figure of a producer element.

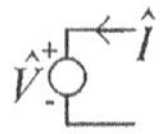

FIGURE 1.4
Figure of a consumer element.

Current flows from the positive terminal of a producer element.
   (*P*) is production active power and (–*P*) is consumption active power.
   (*Q*) is production reactive power and (–*Q*) is consumption reactive power
(Figure 1.3).
   Current enters the positive terminal of a consumer element.
   (*P*) is consumption active power and (–*P*) is production active power.
   (*Q*) is consumption reactive power and (–*Q*) is production reactive power
(Figure 1.4).
   It should be noted that every producer is also a consumer, that is, it con-
sumes in proportion to its negative production, and vice versa.
   Generators are producers with $P>0$ and consumers with $P<0$.
   Motors are consumers with $P>0$ and producers with $P<0$. Note: In prac-
tice, we work with positive sign.

1.7 Types of Load Models

1.7.1 An Impedance Load or a Constant Impedance Load

They are modeled with $Z=R+jX$ like an incandescent lamp or heater.
   The relationship between active power and frequency:

$$\frac{\Delta P}{P} = -2(\sin\varphi)^2 \times \frac{\Delta f}{f} \tag{1.11}$$

The relationship between active power and voltage:

$$\frac{\Delta P}{P} = 2\frac{\Delta V}{V} \tag{1.12}$$

Voltage and frequency change together, so the aforementioned relationships
add up.

1.7.2 A Constant Power Load

They are modeled with $S = P + jQ$ like an AC induction motor.
The relationship between active power and frequency:
Power consumption increases with increasing frequency and vice versa.
The relationship between active power and voltage:
The amount is almost constant and does not change.

1.8 Power and Impedance Triangles

The following can be done to convert a constant impedance load to a constant power load, and vice versa (Figures 1.5 and 1.6):

$$P = V \cdot I \cdot \cos(\varphi) = R \cdot I^2, Q = V \cdot I \cdot \sin(\varphi) = X \cdot I^2, S = \hat{V} \cdot \hat{I}^* = Z \cdot I^2 = \frac{V^2}{Z^*} \quad (1.13)$$

The power triangle is obtained by multiplying the impedance triangle by I^2.

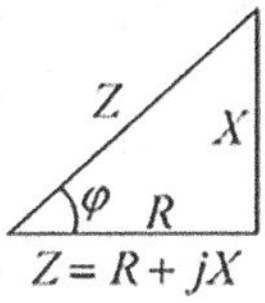

$$Z = R + jX$$

FIGURE 1.5
Impedance triangle.

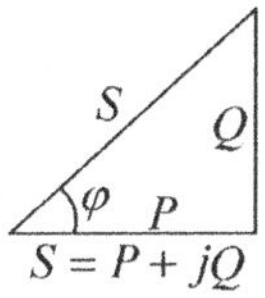

$$S = P + jQ$$

FIGURE 1.6
Power triangle.

1.9 Y-Δ, Star-Delta Transform

Symmetric Star-Delta (Y-Δ) conversion:

$$Z_\Delta = 3 \cdot Z_Y \quad (1.14)$$

1.10 Three-Phase Systems

Instantaneous power in three asymmetric phases:

$$P_{3\text{ph}}(t) = v_a(t) \cdot i_a(t) + v_b(t) \cdot i_b(t) + v_c(t) \cdot i_c(t) \qquad (1.15)$$

A balanced three-phase voltage (equal magnitude and ±120° phase shift) - time display:

$$v_a(t) = \sqrt{2}V \cdot \cos(\omega t + \theta_v)$$

$$v_b(t) = \sqrt{2}V \cdot \cos(\omega t + \theta_v - 120) \qquad (1.16)$$

$$v_c(t) = \sqrt{2}V \cdot \cos(\omega t + \theta_v + 120)$$

A balanced three-phase current (equal magnitude and ±120° phase shift) - time display:

$$i_a(t) = \sqrt{2}I \cdot \cos(\omega t + \theta_i)$$

$$i_b(t) = \sqrt{2}I \cdot \cos(\omega t + \theta_i - 120) \qquad (1.17)$$

$$i_c(t) = \sqrt{2}I \cdot \cos(\omega t + \theta_i + 120)$$

A balanced three-phase voltages and currents, phasor representation:

$$\hat{V}_a = V\angle\theta_v, \quad \hat{V}_b = V\angle(\theta_v - 120), \quad \hat{V}_c = V\angle(\theta_v + 120) \qquad (1.18)$$

$$\hat{I}_a = I\angle\theta_i, \quad \hat{I}_b = I\angle(\theta_i - 120), \quad \hat{I}_c = I\angle(\theta_i + 120)$$

Line-to-line voltage or line voltage:

$$\hat{V}_{ab} = \hat{V}_a - \hat{V}_b = \sqrt{3}V\angle(\theta_v + 30) = V_L\angle(\theta_v + 30)$$

$$\hat{V}_{bc} = \hat{V}_b - \hat{V}_c = \sqrt{3}V\angle(\theta_v - 90) = V_L\angle(\theta_v - 90) \qquad (1.19)$$

$$\hat{V}_{ca} = \hat{V}_c - \hat{V}_a = \sqrt{3}V\angle(\theta_v + 150) = V_L\angle(\theta_v + 150)$$

For balanced three-phase voltages:

$$\hat{V}_a + \hat{V}_b + \hat{V}_c = 0, \qquad \hat{V}_{ab} + \hat{V}_{bc} + \hat{V}_{ca} = 0 \tag{1.20}$$

The angle between voltage and current:

$$\varphi = \theta_v - \theta_i \tag{1.21}$$

The active power in a balanced three-phase system is:

$$P_{3\,ph} = P_{av} = \frac{1}{T}\int P_{3\,ph}(t)\cdot dt = 3V\cdot I\cdot\cos(\varphi) = \sqrt{3}V_L\cdot I\cdot\cos(\varphi) \tag{1.22}$$

The reactive power in a balanced three-phase system is:

$$Q_{3\,ph} = 3VI\cdot\sin(\varphi) = \sqrt{3}\,V_L I\cdot\sin(\varphi) \tag{1.23}$$

The complex power in a balanced three-phase system is:

$$S_{3\,ph} = P_{3\,ph} + jQ_{3\,ph} = 3\cdot\hat{V}\cdot\hat{I}^* = 3\cdot\frac{V^2}{Z^*} \tag{1.24}$$

The apparent power in a balanced three-phase system is:

$$\left|S_{3\,ph}\right| = 3V\cdot I = \sqrt{3}V_L\cdot I = 3\cdot\frac{V^2}{|Z|}$$

1.11 Power Factor Correction

Consider two reactive powers (Q_1 and Q_2) in the power triangle in Figure 1.7, with the real power (P) being constant. As a result, you have two angles (φ_1) and (φ_2). The Q_C shunt capacitor will reduce the reactive power consumed by the load from Q_1 to Q_2. Reactive powers Q_1 and Q_2 are usually replaced by active power P and power factors PF_1 and PF_2.

$$Q_C = Q_1 - Q_2 = P\big(\tan(\varphi_1) - \tan(\varphi_2)\big) \tag{1.25}$$

FIGURE 1.7
Power triangle with two reactive powers Q1 and Q2.

The capacitor impedance and capacitor size in stars are equal to:

$$X_C = \frac{V^2}{Q_C}, \quad C = \frac{Q_C}{2\pi f V^2} \tag{1.26}$$

In a single-phase system, V is the phase voltage and Q_C is the reactive power of one phase, and in a three-phase system, V is the line voltage and Q_C is the reactive power of three phases.

1.12 Per-Unit (p.u.)

Parameters are per-unit by dividing them by parameter bases. Parameters and base parameters:

$$\underbrace{S(VA), P(W), Q(VAr)}_{S_b (VA)}, \quad \underbrace{Z(\Omega), R(\Omega), X(\Omega)}_{Z_b (\Omega)}, \quad \underbrace{Y(\mho), G(\mho), B(\mho)}_{Y_b (\mho)}, \quad \underbrace{V(V)}_{V_b (V)}, \quad \underbrace{I(A)}_{I_b (A)} \tag{1.27}$$

1.12.1 Per-Unit System for Single-Phase

V_b phase voltage and S_b single-phase power are generally the two main bases used. All other bases can be calculated from the following equations. The nominal values are usually equal to these values. It is usually assumed that the system's maximum rated power is S_b. In the first generator, the voltage is V_b. This voltage is passed through the transformers with a conversion ratio.

$$Z_b = \frac{V_b^2}{S_b}, \quad I_b = \frac{V_b}{Z_b}, \quad Y_b = \frac{1}{Z_b} \tag{1.28}$$

1.12.2 Per-Unit System for Three-Phase

S_b three-phase power and V_b line-to-line voltage are usually known as the main bases. All other bases can be calculated from the following equations:

$$Z_b = \frac{V_b^2}{S_b}, \quad I_b = \frac{V_b}{\sqrt{3}.Z_b}, \quad Y_b = \frac{1}{Z_b} \qquad (1.29)$$

Three-phase per-unit systems eliminate coefficients 3 and $\sqrt{3}$ in power and voltage, such as $P = V \cdot I \cdot \cos(\varphi)$ for three-phase power.

1.12.3 Base Changes in the Per-Unit System

$$Z_{new}\,(p.u.) = Z_{old}\,(p.u.) \cdot \frac{S_b^{new}}{S_b^{old}} \cdot \left(\frac{V_b^{old}}{V_b^{new}} \right)^2 \qquad (1.30)$$

1.13 Transmission Power

In the aforementioned system (Figure 1.8):
S_{gi} is the generator production power in the ith bus;
S_{di} is the load demand power in the ith bus;
S_i is the complex power injected into the ith bus;

$$\text{And} \quad S_i = S_{gi} - S_{di}$$

The phase difference between V_1 and V_2:

$$\delta = \delta_{12} = \delta_1 - \delta_2 \qquad (1.31)$$

Line current from bus 1 to bus 2:

$$\hat{I} = \hat{I}_{12} = \frac{\hat{V}_1 - \hat{V}_2}{Z} \qquad (1.32)$$

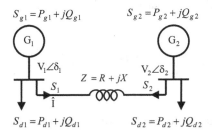

FIGURE 1.8
A two-bus network.

The complex power injected into the bus:

$$S_1 = \hat{V}_1 \times \hat{I}^* = P_1 + jQ_1, \quad S_2 = \hat{V}_2 \times \hat{I}^* = P_2 + jQ_2 \tag{1.33}$$

The active power injected into bus 1:

$$P_1 = \frac{R\left(V_1^2 - V_1 V_2 \times \cos(\delta)\right)}{R^2 + X^2} + \frac{X V_1 V_2 \times \sin(\delta)}{R^2 + X^2} \tag{1.34}$$

The reactive power injected into bus 1:

$$Q_1 = \frac{X\left(V_1^2 - V_1 V_2 \times \cos(\delta)\right)}{R^2 + X^2} - \frac{R V_1 V_2 \times \sin(\delta)}{R^2 + X^2} \tag{1.35}$$

You can determine the injection power of bus 2 by converting 1 in the above equation to 2 and vice versa (δ becomes $-\delta$).

The active power injected into bus 2:

$$P_2 = \frac{R\left(V_2^2 - V_1 V_2 \times \cos(\delta)\right)}{R^2 + X^2} - \frac{X V_1 V_2 \times \sin(\delta)}{R^2 + X^2} \tag{1.36}$$

The reactive power injected into bus 2:

$$Q_2 = \frac{X\left(V_2^2 - V_1 V_2 \times \cos(\delta)\right)}{R^2 + X^2} + \frac{R V_1 V_2 \times \sin(\delta)}{R^2 + X^2} \tag{1.37}$$

1.13.1 First Special Conditions ($R = 0$)

$$P_1 = \frac{V_1 V_2}{X} \sin(\delta), \quad P_2 = -\frac{V_1 V_2}{X} \sin(\delta) \tag{1.38}$$

$$Q_1 = \frac{V_1^2 - V_1 V_2 \times \cos(\delta)}{X}, \quad Q_2 = \frac{V_2^2 - V_1 V_2 \times \cos(\delta)}{X} \tag{1.39}$$

Active and reactive power consumption in the line:

$$Q_{line} = Q_1 + Q_2 = X \times I^2, \quad P_{line} = P_1 + P_2 = 0 \tag{1.40}$$

The average of reactive power on the line:

$$Q_{av} = \frac{Q_1 - Q_2}{2} = \frac{V_1^2 - V_2^2}{2X} \tag{1.41}$$

1.13.2 Second Special Conditions ($R = 0$, $V_1 = V_2 = V$)

$$P = P_1 = -P_2 = \frac{V^2}{X} \sin(\delta) \tag{1.42}$$

$$Q = Q_1 = Q_2 = \frac{Q_{line}}{2} = \frac{V^2(1 - \cos(\delta))}{X} \tag{1.43}$$

In a stable power system: $0 \le \delta \le 90°$

$$P \ge Q = Q_G - Q_D \ge 0 \tag{1.44}$$

- In other words, both buses produce reactive power.
- This line consumes reactive power.
- The reactive power produced by each bus is greater than the reactive power consumed by the same bus.
- In the line, the active power injected is greater than the reactive power injected.

1.13.3 Double-Sided Feeding Line, or Line Connecting Both Ends of the Power Supply

When both V_1 and V_2 are constant, the line is called a double-sided feeding line. Any reactive power controller, such as a capacitor, generator, or synchronous capacitor (condenser), will keep the line voltage constant.

The maximum active power is calculated as follows:

$$R \ne 0 \Rightarrow \delta = \tan^{-1}\left(\frac{X}{R}\right) \Rightarrow P_2 \text{ is maximum} \tag{1.45}$$

$$R = 0 \Rightarrow \delta = 90° \Rightarrow P_1 = -P_2 \text{ is maximum} = \frac{V_1 V_2}{X} \tag{1.46}$$

1.13.4 One-Sided Feeding Line or Radial Transmission Line

When the bus voltage $2(V_2)$ varies according to the load, the line is called a one-sided feeding line.

It is possible that bus 2 has a reactive power controller, but it is out of control. In this case, we have $Q_2=0$ in relation to equation (1.39) as follows:

$$V_1 \cos\delta = V_2 \Rightarrow P_1 = \frac{V_1 V_2}{X} \sin\delta \Rightarrow P_1 = \frac{V_1^2 \sin\delta \cos\delta}{X} \Rightarrow P_1 = \frac{V_1^2}{2X} \sin 2\delta \qquad (1.47)$$

$$R = 0 \Rightarrow \delta = 45° \Rightarrow P_1 \text{ is maximum} = \frac{V_1^2}{2X} \qquad (1.48)$$

1.14 Generator and Motor Connected

A source that generates positive active power is a generator, and one that consumes positive active power is a motor. Their reactive power has nothing to do with their active power.

1.14.1 Generator and Motor Relations – Similar to Double-Sided Feeding Line, First Special Conditions ($R = 0$)

V_1 is a generator and V_2 is a motor if:

$$\delta > 0 \Rightarrow \delta_1 > \delta_2 \qquad (1.49)$$

V_1 produces reactive power if:

$$V_1 > V_2 \cdot \cos(\delta) \qquad (1.50)$$

V_2 produces reactive power if:

$$V_2 > V_1 \cdot \cos(\delta) \qquad (1.51)$$

Both V_1 and V_2 cannot be motors or generators at the same time.

1.14.2 Generator and Motor Relations – Similar to Double-Sided Feeding Line, Second Special Conditions ($X = 0$)

$$P_1 = \frac{V_1^2 - V_1 V_2 \cdot \cos(\delta)}{R}, \quad \text{The active power injected into bus 1} \quad (1.52)$$

$$P_2 = \frac{V_2^2 - V_1 V_2 \cdot \cos(\delta)}{R}, \quad \text{The active power injected into bus 2} \quad (1.53)$$

$$Q_1 = -\frac{V_1 V_2}{R} \sin(\delta), \quad \text{The reactive power injected into bus 1} \quad (1.54)$$

$$Q_2 = \frac{V_1 V_2}{R} \sin(\delta), \quad \text{The reactive power injected into bus 2} \quad (1.55)$$

$$V_1 > V_2 \cdot \cos(\delta), \quad \text{If } V_1 \text{ is a generator} \quad (1.56)$$

$$V_2 > V_1 \cdot \cos(\delta), \quad \text{If } V_2 \text{ is a generator} \quad (1.57)$$

$$\delta > 0 \Rightarrow \delta_1 > \delta_2, \quad V_2 \text{ produces and } V_1 \text{ consumes reactive power} \quad (1.58)$$

1.15 Important Numerical Approximations

$$\sin(37°) = \cos(53°) = 0.6$$
$$\cos(37°) = \sin(53°) = 0.8$$
$$\tan(37°) = 0.75 = \frac{3}{4}$$
$$\tan(53°) = \frac{4}{3}$$
$$1\angle 0 + 1\angle 60° = \sqrt{3}\angle 30°$$
$$1\angle 60° - 1\angle 0 = 1\angle 120°$$

$$(1.59)$$

Part Two: Answer Question

1.16 Four-Choice Questions – 47 Questions

1.1. What is the production reactive power of bus 2 in Figure 1.9 system?

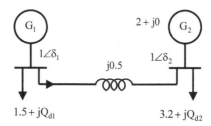

FIGURE 1.9
Question network 1.1.

    1. −0.6   2. 0.6   3. −0.4   4. 0.4

1.2. In question 1.1, what is the reactive power loss?

    1. 0.8   2. −0.8   3. 0   4. 0.6

1.3. In the absence of the neutral of the generator, what is the short circuit current from one phase to the ground of a balanced three-phase source of 10 A?

    1. $10\angle0$ 2. $20\angle120$ 3. 0   4. $10\angle0$

1.4. Using a three-phase line with a voltage of 220 V, a triangle with impedance $Z = 30\Omega + j0$ in every phase is connected. Find the line-to-line voltage of the source terminal if the transmission line impedance is $Z_1 = j10\Omega$.

    1. $220\left(1 + j\sqrt{3}\right)$ 2. $220(3 + j)$

    3. $\dfrac{220\sqrt{3}}{3}(1 + j)$ 4. $220(1 + j)$

1.5. A single-phase voltage source of 220 V is connected in parallel with two impedances and a single-phase generator. How much can this resource produce all at once? (cos 36=0.8)

$$Z_1 = 4840\Omega\angle36°, \quad Z_2 = 9680\Omega\angle-53°, \quad G:5\,\text{VA}, 0.8\,\text{lead}$$

    1. $7 - j5$ 2. $15 - j$ 3. $7 + j5$ 4. $-7 - j5$

1.6. A generator is connected to an infinite network under the following conditions. The active output power of the generator remains constant, but its excitation is reduced by 50%. What is the percentage of reactive changes in production?

$$X_S = 1\text{p.u}, \ V_t = 1\text{p.u}, \ I = 1\text{p.u}, \ PF = \frac{\sqrt{3}}{2} \text{ lag}$$

1. 50% increase   2. 25% reduction

3. 50% reduction   4. 250% increase

1.7. A 5 Ω resistor and a 5 Ω capacitor are in series in a single-phase load. How much of the reactive power changes when both voltage and frequency drop by 1% simultaneously?

1. 1% increase   2. 1% reduction

3. 2% increase   4. 2% reduction

1.8. What is the maximum transfer power between the generator and the load in Figure 1.10 system if the load voltage is constant?

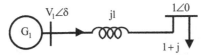

FIGURE 1.10
Question network 1.8.

1. $\sqrt{5}$ 2. $10\sqrt{5}$ 3. 10   4. $5\sqrt{10}$

1.9. What is the maximum transmission power if the load voltage is not constant and $Q_{d2} = 0$ in question **1.8**?

1. 1   2. 0.5   3. Cannot be calculated   4. $\sqrt{2}$

1.10. The resistance of a DC transmission line is R and the average transmitted power is P_{av}. The average reactive power of an AC line with a reactance of X ($R=0$) is Q_{av}. What is the ratio of P_{av} to Q_{av} if the DC voltage is equal to the RMS AC voltage?

1. $\dfrac{R}{X}$ 2. $\dfrac{V_1^2 - V_2^2}{2RX}$ 3. $\dfrac{X}{R}$ 4. $\dfrac{V_1^2 - V_2^2}{4RX}$

1.11. Find the updated reactance of transformer 1 with the following base values in Figure 1.11 single-line diagram of a three-phase power system?

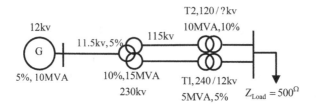

FIGURE 1.11
Question network 1.11.

$$S_b = 10\,\text{MVA} \quad V_{bG} = 12 \text{ kV}$$

1. 5%   2. 12%   3. 10%   4. 2%

1.12. Which is the average reactive power on the line and the reactive power consumption of the line, respectively, in Figure 1.12 system?

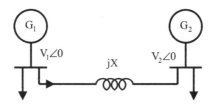

FIGURE 1.12
Question network 1.12.

1. $\dfrac{1}{2X}\left(V_1^2 + V_2^2\right),\ \dfrac{1}{X}\left(V_1^2 - V_2^2\right)$ 2. $\dfrac{1}{X}\left(V_1^2 - V_2^2\right),\ \dfrac{1}{2X}\left(V_1^2 + V_2^2\right)$

3. $\dfrac{1}{X}\left(V_1^2 + V_2^2\right),\ \dfrac{1}{X}\left(V_1^2 + V_2^2\right)$ 4. $\dfrac{1}{2X}\left(V_1^2 - V_2^2\right),\ \dfrac{1}{X}\left(V_1 - V_2\right)^2$

1.13. In Figure 1.13, what is the reactive power of the capacitor in bus 3?

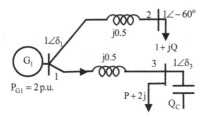

FIGURE 1.13
Question network 1.13.

    1. 2   2. $-\sqrt{3}$ 3. $4-\sqrt{3}$ 4. $3\sqrt{3}$

1.14. Capacitor bus 2 can be controlled in Figure 1.14 system. What is the magnitude and phase voltage of bus 2 when the capacitor is connected to bus 2 compared to when the capacitor is disconnected?

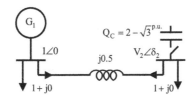

FIGURE 1.14
Question network 1.14.

    1. 1, 1   2. $\dfrac{\sqrt{2}}{2}, \dfrac{3}{2}$ 3. $\sqrt{2}, \dfrac{2}{3}$ 4. $\sqrt{2}, \dfrac{3}{2}$

1.15. In Figure 1.15 network, two generators have the same output power. When $S_d = 200\,\text{MW}$, the losses of lines 1 and 2 are each 1 MW. At load $S_d = 400\,\text{MW}$, what is the total system loss?

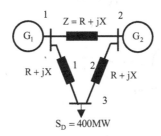

FIGURE 1.15
Question network 1.15.

    1. 2 MW   2. 4 MW   3. 16 MW   4. 8 MW

1.16. Connect two electric machines with the following voltages and one impedance. Which machine works as a generator and which one as a motor?

$$\hat{E}_1 = 100\angle30, \quad \hat{E}_2 = 120\angle0, \quad Z = 2 + j5\Omega$$

1. E_1 motor, E_2 generator
2. E_2 motor, E_1 generator
3. E_1 motor, E_2 motor
4. E_1 generator, E_2 generator

1.17. What is the line current in Figure 1.16?

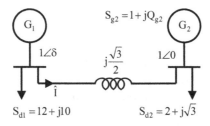

FIGURE 1.16
Question network 1.17.

1. $\dfrac{\sqrt{3}}{4}\angle60$
2. $\dfrac{2}{\sqrt{3}}\angle30$
3. $\sqrt{3}\angle30$
4. $\dfrac{\sqrt{3}}{2}\angle30$

1.18. In question 1.17, what is Q_{g2} if $V_2 = 1$?

1. $\dfrac{2}{\sqrt{3}}$
2. $-\dfrac{2}{\sqrt{3}}$
3. $\dfrac{4}{\sqrt{3}}$
4. $\dfrac{1}{\sqrt{3}}$

1.19. Assume δ is $<60°$ for preventing static instability in question **1.17**. What is the maximum transmission power for the line?

1. $\dfrac{2}{\sqrt{3}}$
2. 1
3. $\dfrac{\sqrt{3}}{2}$
4. $\dfrac{1}{2}$

1.20. Calculate the line reactance value per-unit for the single-line diagram in Figure 1.17.

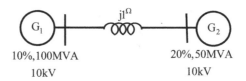

FIGURE 1.17
Question network 1.20.

1. 0.5 p.u.
2. 1 p.u.
3. 1 or 0.5 p.u.
4. Cannot be specified

1.21. The three-phase transmission line feeds a star load at its end with impedance $\left(Z_S = 0.5 + j2\,\Omega\right)$ and 60 Hz frequency. What capacitors do we need in each phase if we want the load power factor to be 0.8? (Q_C, Z_C) (Load: 240MW, 10kV, 0.6 lag)

1. $\dfrac{7}{5}\Omega$, 240 MVAr   2. $\dfrac{5}{7}\Omega$, 140 MVAr

3. $\dfrac{15}{7}\Omega$, $\dfrac{140}{3}$ MVAr   4. $\dfrac{5}{21}\Omega$, 420 MVAr

1.22. As an example, consider a three-phase line with $2\sqrt{3}$ kV and 90 kVA. The voltage drop across the resistor and the line reactance, respectively, are $1/\sqrt{3}\,\%$ and $\sqrt{3}\,\%$ of the rated voltage. With a power factor of 0.8 lag and a voltage of $2\sqrt{3}$ kV, the line supplies a load of 48 kW. Determine the power consumption of the line.

1. $1200 + j400$ 2. $400 + j1200$

3. $480 + j1440$ 4. $1440 + j480$

1.23. Calculate $(Z\ \text{p.u.})$ and $(Y_2\ \text{p.u.})$ in Figure 1.18 system.

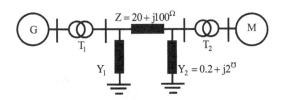

FIGURE 1.18
Question network 1.23.

	X(p.u)	$V_{(n)}$(kV)	$S_{(n)}$(MVA)
G	0.1	20	100
M	0.2	30	90
T_1	0.1	14/140	80
T_2	0.1	140/30	90

1.
$Z = 0.5 + j2.5$
$Y_2 = 8 + j8$

2.
$Z = 0.2 + j10$
$Y_2 = 0.2 + j2$

3.
$Z = 0.02 + j0.1$
$Y_2 = 20 + j200$

4.
$Z = 0.05 + j0.25$
$Y_2 = 80 + j800$

1.24. Through a series impedance transmission line, machines 1 and 2 are connected. IF: $\hat{E}_1 = 200\angle -30°$, $\hat{E}_2 = 200\angle 0°$, $Z = j5\Omega$

  1. Machine E_1 *produces* reactive power, whereas machine E_2 *consumes* reactive power. Machine E_1 works as a *generator*, whereas machine E_2 works as a *motor*.

  2. Machine E_1 *consumes* reactive power, whereas machine E_2 *produces* reactive power. Machine E_1 works as a *motor*, whereas machine E_2 works as a *generator*.

  3. Machines E_1 and E_2 *produce* reactive power equally. Machine E_1 works as a *generator*, whereas machine E_2 works as a *motor*.

  4. Machines E_1 and E_2 *produce* reactive power equally. Machine E_1 works as a *motor*, whereas machine E_2 works as a *generator*.

1.25. In a power system, 100 MVA and 20 kV are the base values. What is the reactor reactance in p.u. with characteristics $(200\,\text{MVAr}, 20\,\text{kV})$?

  1. 0.25   2. 0.75   3. 0.5   4. 2

1.26. If the voltage of the buses is 1 p.u., what should the reactive power produced by the capacitor (Q_{C2}) be in terms of per-unit (Figure 1.19)?

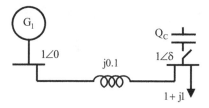

FIGURE 1.19
Question network 1.26.

Suppose: $\cos\left(\sin^{-1}(0.1)\right) = 0.995$

  1. 1.05   2. 0.95   3. 1.15   4. 0.05

1.27. There are three loads ($P=2$, 0.8 lag), ($P=2$, 0.8 lead) and ($P=2$, PF=1) connected to a bus with voltage ($1\angle -12$). The total load can be shown as equivalent admittance. What is the p.u. size of this admittance?

  1. 6   2. 2–j   3. 2+j   4. 2–2j

1.28. A capacitor bank at the end of a 400 V and 50 Hz three-phase line consists of three capacitors connected in a triangle and delivering 600 kVAr to the system. What is the capacity of each capacitor?

  1. 5000 μF   2. 4000 μF   3. 0.004 μF   4. 0.005 F

1.29. In Figure 1.20 system, the delta transmission angle $(\delta = \delta_1 - \delta_2)$ is 15°. How does the magnitude of the current I and its angle with respect to E_2 change if this angle increases slightly without changing the magnitude of the voltage? (Current I always lags E_2.)

FIGURE 1.20
Question network 1.29.

1. The magnitude of (I) *increases*, and its angle with respect to E_2 *increases*.

2. The magnitude of (I) *decreases*, and its angle with respect to E_2 *decreases*.

3. The magnitude of (I) *increases*, and its angle with respect to E_2 *decreases*.

4. The magnitude of (I) *decreases*, and its angle with respect to E_2 *increases*.

1.30. What are the values of P and Q in Figure 1.21 system in terms of p.u.?

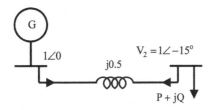

FIGURE 1.21
Question network 1.30.

$$\sin(15°) = 0.25, \quad \cos(15°) = 0.96$$

1. $P=0.5$ p.u., $Q=0.08$ p.u.   2. $Q=0.5$ p.u., $P=0.8$ p.u.
3. $Q=-0.5$ p.u., $P=0.8$ p.u.   4. $P=0.5$ p.u., $Q=-0.08$ p.u.

1.31. What is the amount of load resistance per-unit in the system of
Figure 1.22 if the base values are equal to the nominal values of the
generator?

G: 20 kV, 300 MVA; T_1: 20/200 kV, 375 MVA; T_2: 180/9 kV, 300 MVA;
Load: 9 kV, 180 MVA

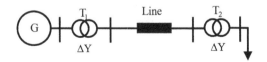

FIGURE 1.22
Question network 1.31.

    1. 1.25   2. 1.35   3. 1.45   4. 1.55

1.32. At an impedance of one with a power factor of 0.8, what is the change
in the active power consumption if the frequency increases by 10%?

    1. 3.6% increase   2. 7.2% decrease

    3. 7.2% increase   4. 3.6% decrease

1.33. A voltage source is connected to a load that has an impedance of
1 p.u. and a power factor of 0.8 lag. How much does the complex
power consumption of the load increase when the source frequency
is doubled?

    1. It increases 0.5 times   2. Almost constant

    3. It increases $\sqrt{2}$ times   4. It increases $(2)^{-0.5}$ times

1.34. When the voltage is doubled in one impedance load, how much is
the power consumption at a constant frequency multiplied?

    1. 0.25   2. 4   3. 2   4. 0.5

1.35. What is the delta angle (δ) at which the maximum active power is
transmitted from bus 1 to bus 2 in Figure 1.23 system?

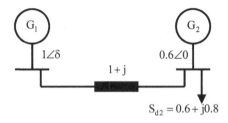

FIGURE 1.23
Question network 1.35.

    1. 36   2. 54   3. 45   4. 30

1.36. The instantaneous power in a single phase with a sinusoidal voltage source is as follows. Which of the following represents the complex power delivery per load?

$$v(t) = 200\cos(2t), \phi_0 > 0, p(t) = 800 + 1000\cos(4t - \phi_0)$$

1. $800 + j600$ 2. $800 - j600$

3. $600 - j800$ 4. $600 + j800$

1.37. 1.2 kW of active power is consumed by a single-phase induction load with a power factor of 0.6 lag. If the source voltage is 200 V, what are R and X (parallel)?

1. $R = 14, X = 25$ 2. $R = 12, X = 16$

3. $R = 33.3, X = 25$ 4. $R = 33.3, X = 16$

1.38. A three-phase circuit should have a certain power factor. What is the reactance of each unit of capacitor in the triangle state compared with the star state if we close the capacitors in a triangle and a star form?

1 $\sqrt{3}$ 2. $\dfrac{1}{\sqrt{3}}$ 3. $\dfrac{1}{3}$ 4. 3

1.39. Generator (10 kV, 50 HZ) connected to a resistive load ($R = 1\ \Omega$) via line ($Z = j1$). What is the capacity of the capacitor added to the load if we want the resistance voltage (R) to be 10 kV? ($\pi = 3$)

1. 3600 µF   2. 3300 µF   3. 6800 µF   4. 4000 µF

1.40. The magnitude of V_2 always becomes larger than V_1. What angle of current is present on the following transmission line with a leading load (Figure 1.24)? (Generator exciter keeps voltage V_1 constant.)

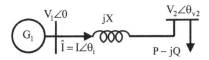

FIGURE 1.24
Question network 1.40.

1. $\theta_i < 0$ 2. $\theta_i > 0$ 3. $\theta_{v2} > \theta_i > 0$

4. In the leading load, we always have: $V_2 > V_1$

1.41. Which option is correct for the leading load and Figure 1.25 transmission line?

FIGURE 1.25
Question network 1.41.

1. For some loads, V_2 is greater than V_1, and for others, it is smaller than V_1.
2. V_1 is always smaller than V_2.
3. V_1 is always larger than V_2.
4. V_2 becomes greater than V_1 if the current angle is smaller than the angle (δ).

1.42. What is V2 in the following system? (Double-sided feeding line) (Figure 1.26)

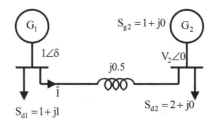

FIGURE 1.26
Question network 1.42.

　　1. $\sqrt{0.4}$　　　　　2. 0.6　　　　　3. $\sqrt{0.5}$　　　　　4. 0.808

1.43. How much reactive power is generated by generator 1 in Question 1.42 and what is the line current?

　　1. $Q_{g1} = 1, I = 1$　　　　　　　　2. $Q_{g1} = 2, I = \sqrt{2}$

　　3. $Q_{g1} = 2, I = 1$　　　　　　　　4. $Q_{g1} = 1, I = 0.717$

1.44. In the case of constant bus voltage, for which δ does the second bus consume reactive power (Figure 1.27)?

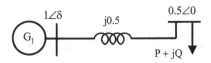

FIGURE 1.27
Question network 1.44.

    1. $\delta > 60$ 2. $\delta < 60$ 3. Always   4. Never

1.45. What is the bus voltage (1) (Figure 1.28)?

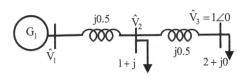

FIGURE 1.28
Question network 1.45.

    1. $1 + j1.5$ 2. $2 + j1.5$ 3. $1 + j2$ 4. $1 + j2.5$

1.46. According to Figure 1.29 network, if generator (G_1) is out of service, the Thevenin admittance of this bus is equal to $-j_2$ p.u. Conversely, if generator (G_2) is out of service, the Thevenin admittance is equal to $-j1$ p.u. What is the value of reactance (X)?

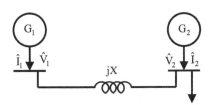

FIGURE 1.29
Question network 1.46.

    1. 2 p.u.   2. 1.5 p.u.   3. 1 p.u.   4. 0.5 p.u.

1.47. Resistance and capacitance are ignored in Figure 1.30 network, and series impedance is equal. What is the correct statement if Q_C is a reactive power compensator? $|V_1| = |V_2| = |V_3|$

FIGURE 1.30
Question network 1.47.

1. In the line between 1 and 3, there is no reactive power.
2. In the line between 2 and 3, there is no reactive power.
3. Reactive power generated by the generator (1) accounts for half of the power consumed between 2 and 3.
4. Reactive power generated by the generator (1) accounts for half of the power consumed between 1 and 3.

1.17 Key Answers to Four-Choice Questions

Question	1	2	3	4
1. (4)				×
2. (1)	×			
3. (3)			×	
4. (4)				×
5. (3)			×	
6. (3)			×	
7. (2)		×		
8. (1)	×			
9. (1)	×			
10. (3)			×	
11. (3)			×	
12. (4)				×
13. (3)			×	
14. (3)			×	
15. (4)				×
16. (2)		×		
17. (2)		×		
18. (3)			×	

(Continued)

Question	1	2	3	4
19. (2)		×		
20. (2)		×		
21. (2)		×		
22. (2)		×		
23. (4)				×
24. (4)				×
25. (3)			×	
26. (1)	×			
27. (1)	×			
28. (2)		×		
29. (3)			×	
30. (4)				×
31. (2)		×		
32. (2)		×		
33. (4)				×
34. (2)		×		
35. (3)			×	
36. (1)	×			
37. (3)			×	
38. (4)				×
39. (2)		×		
40. (2)		×		
41. (3)			×	
42. (3)			×	
43. (2)		×		
44. (2)		×		
45. (4)				×
46. (4)				×
47. (4)				×

1.18 Descriptive Answers to Four-Choice Questions

1.1. **Option 4 is correct.** From equations (1.38) and (1.39), we have:

$$P_{12} = \frac{V_1 V_2}{0.5} \sin \delta = (3.2 - 2) = \frac{1 \times 1}{0.5} \sin \delta \Rightarrow \sin \delta = 0.6$$

$$Q_2 = \frac{V_2^2 - V_1 V_2 \cdot \cos \delta}{X} = \frac{1 - 0.8}{0.5} = 0.4$$

1.2. **Option 1 is correct.** According to the answer to question 1.1, we have:

$$Q_1 = \frac{V_1^2 - V_1 V_2 \cos \delta}{X} = 0.4$$

From equation (1.40): $Q_{line} = Q_1 + Q_2 = 0.4 + 0.4 = 0.8$

1.3. **Option 3 is correct.** A neutral wire has no current because there is no path to reverse the current on the source side.

1.4. **Option 4 is correct.** By converting the triangle to a star and a simple KVL, we have:

$$\hat{I} = \frac{220}{\sqrt{3} \times 10} \Rightarrow \hat{V}_{ph} = \hat{I} \times (j10) + \frac{220}{\sqrt{3}} = \frac{\sqrt{3}}{3}(220 + j220) \Rightarrow \hat{V}_L = 220 + j220$$

1.5. **Option 3 is correct.** From equation (1.13), we have:

$$\text{Consumption:} \quad S_1 = \frac{V^2}{Z_1^*} = \frac{(220)^2}{4840 \angle -36} = 10 \angle 36 = 8 + j6$$

$$\text{Consumption:} \quad S_2 = \frac{V^2}{Z_2^*} = \frac{(220)^2}{9680 \angle +53} = 5 \angle -53 = 3 - j4$$

$$\text{Production:} \quad S_G = 5 \times 0.8 - j5 \times 0.6 = 4 - j3$$

$$S_{Source} = S_1 + S_2 - S_G \Rightarrow S = 11 + j2 - (4 - j3) = 7 + j5$$

1.6. **Option 3 is correct.** Displayed before increasing with (old) and after increasing with (new).

$$\hat{I} = 1 \angle -\cos^{-1} \frac{\sqrt{3}}{2} = 1 \angle -30°$$

$$\hat{E} = j1 \angle -30 + 1 \angle 0 = \sqrt{3} \angle 30$$

$$P_{old} = \frac{E V_t}{X} \sin \delta_{old} = \frac{\sqrt{3} \times 1}{1} \sin(30 - 0) = \frac{\sqrt{3}}{2}$$

$$\text{Production:} \quad Q_{old} = \frac{E^2 - V_t E \cos \delta}{X} = \frac{\left(\sqrt{3}\right)^2 - 1 \times \sqrt{3} \cos 30}{1} = 1.5$$

$$E_{new} = \sqrt{3} - 0.5\sqrt{3} = 0.5\sqrt{3}$$

$$P_{new} = P_{old} = \frac{\sqrt{3}}{2} = \frac{0.5\sqrt{3} \times 1}{1} \sin\delta_{new} \Rightarrow \sin\delta_{new} = 1 \Rightarrow \cos\delta_{new} = 0$$

$$Q_{new} = \frac{E^2 - V_t \cdot E\cos\delta}{X} = \frac{\left(0.5\sqrt{3}\right)^2 - 0}{1} = 0.75$$

$$\%\Delta Q = \frac{Q_2 - Q_1}{Q_1} = \frac{0.75 - 1.5}{1.5} \times 100 = -50$$

1.7. **Option 2 is correct.** From equations (1.11) and (1.12) we have:

$$\frac{\Delta P}{P} = 2\frac{\Delta V}{V} - 2\left(\sin(\varphi)\right)^2 \frac{\Delta f}{f} = 2\frac{-1}{100} - 2\left(\sin(45°)\right)^2 \frac{-1}{100}$$

$$= \frac{-2}{100} - 2\left(\frac{\sqrt{2}}{2}\right)^2 \times \frac{-1}{100} = \frac{-2}{100} + \frac{1}{100} = \frac{-1}{100} = -1\%$$

1.8. **Option 1 is correct.** From equation (1.46), we have:

$$\hat{I} = \frac{S_{d2}^*}{\hat{V}_2^*} = \frac{1-j}{1} = 1 - j \Rightarrow V_1 = |1\angle 0 + j(1-j)| = |2 + j| = \sqrt{5}$$

$$P_{max} = \frac{V_1 V_2}{X} = \frac{\sqrt{5}}{1} = \sqrt{5}$$

1.9. **Option 1 is correct.** From equation (1.48), we have:

$$S_{d2} = 1 + j0 \Rightarrow \hat{I} = \frac{S_{d2}^*}{\hat{V}_2^*} = \frac{1}{1} = 1\angle 0 \Rightarrow \hat{V}_1 = j1 + 1 = \sqrt{2}\angle 45° \Rightarrow P = \frac{V_1^2}{2X} = \frac{2}{2} = 1$$

1.10. **Option 3 is correct.** From equation (1.41), we have:

$$\left.\begin{aligned} P_{av} &= \frac{V_1^2 - V_2^2}{2R} \\ Q_{av} &= \frac{V_1^2 - V_2^2}{2X} \end{aligned}\right\} \Rightarrow \frac{P_{av}}{Q_{av}} = \frac{X}{R}$$

Note:

$$\left.\begin{aligned} P_1 &= V_1 I = V_1\left(\frac{V_1 - V_2}{R}\right) = \frac{V_1^2 - V_1 V_2}{R} \\ P_{d2} &= V_2 I = V_2\left(\frac{V_1 - V_2}{R}\right) = \frac{V_1 V_2 - V_2^2}{R} \end{aligned}\right\} \Rightarrow P_{av} = \frac{P_1 + P_{d2}}{2} = \frac{V_1^2 - V_2^2}{2R}$$

1.11. **Option 3 is correct.** If V_{b1} is the base voltage of the initial side of transformer T_1:

$$\frac{11.5}{230} = \frac{12\,\text{kV}}{V_{b1}} \Rightarrow V_{b1} = 240\,\text{kV} \Rightarrow \%X_{\text{new}} = 5\%\left(\frac{10\,\text{MVA}}{5\,\text{MVA}}\right)\left(\frac{240}{240}\right)^2 = 10\%$$

1.12. **Option 4 is correct.** From equations (1.40) and (1.41), we have:

$$Q_{av} = \frac{1}{2X}\left(V_1^2 - V_2^2\right)$$

Production: $\quad Q_1 = \frac{1}{X}\left(V_1^2 - V_1 V_2 \cdot \cos\delta\right),\ \ Q_2 = \frac{1}{X}\left(V_2^2 - V_1 V_2 \cdot \cos\delta\right)$

$$Q_{\text{Line}} = Q_1 + Q_2 = \frac{1}{X}\left(V_1^2 + V_2^2 - 2V_1 V_2 \cdot \cos\delta\right) \overset{\delta=0}{=} \frac{\left(V_1 - V_2\right)^2}{X}$$

1.13. **Option 3 is correct.**

The first solution:

$$P_{12} = 1 = \frac{1 \times 1\sin\left(\delta_1 + 60\right)}{0.5} \Rightarrow \sin\left(\delta_1 + 60\right) = \frac{1}{2} \Rightarrow \delta_1 = -30°$$

$$P_{13} = 2 - 1 = \frac{1 \times 1\sin\left(-30 - \delta_3\right)}{0.5} \Rightarrow \sin\left(-30 - \delta_3\right) = 0.5 \Rightarrow \delta_3 = -60°$$

Production: $\quad Q_3 = \dfrac{V_3^2 - V_1 V_3 \cos\left(-30 + 60\right)}{0.5} = \dfrac{1 - \dfrac{\sqrt{3}}{2}}{0.5} = 2 - \sqrt{3}$

$$Q_3 = Q_C - 2 \Rightarrow Q_C = Q_3 + 2 = 2 - \sqrt{3} + 2 = 4 - \sqrt{3}$$

The second solution:

From equation (1.44) should $Q_C > 2$ and $(Q_C - 2) < 1$.

1.14. **Option 3 is correct.**

First mode with a capacitor (stage A)

$$P_{12} = 1 = \frac{1 \times V_2}{0.5}\sin\left(0 - \delta_2\right) \Rightarrow 0.5 = V_2 \cdot \sin\left(-\delta_2\right)$$

Production: $\quad Q_2 = \dfrac{V_2^2 - V_2 \times 1\cos\left(0 - \delta_2\right)}{0.5} = 2\left(V_2^2 - V_2 \cdot \cos\left(\delta_2\right)\right) = 2 - \sqrt{3}$

$$\Rightarrow \quad \left.\begin{array}{l} V_2 = -\dfrac{0.5}{\sin\delta_2} \\[3mm] V_2^2 - V_2 \cos\delta_2 = 1 - \dfrac{\sqrt{3}}{2} \end{array}\right\} \Rightarrow \text{Guess: } V_{2A} = 1,\quad \delta_{2A} = -30$$

Second mode without a capacitor (stage B)

$$P_{12} = 1 = \frac{V_2 \sin(0 - \delta_2)}{0.5} \Rightarrow 0.5 = V_2 \sin(-\delta_2)$$

$$Q_2 = 0 = \frac{V_2^2 - V_2 \cos(0 - \delta_2)}{0.5} \Rightarrow V_2 = \cos(\delta_2)$$

$$0.5 = -\sin(\delta_2)\cos(\delta_2) \Rightarrow 1 = -2\sin\delta_2\cos\delta_2 \Rightarrow \sin 2\delta_2 = -1 \Rightarrow \delta_{2B} = -45°$$

$$\Rightarrow V_{2B} = \frac{\sqrt{2}}{2} \Rightarrow \frac{V_{2A}}{V_{2B}} = \frac{1}{\frac{\sqrt{2}}{2}} = \sqrt{2}, \frac{\delta_{2A}}{\delta_{2B}} = \frac{-30}{-45} = \frac{2}{3}$$

1.15. **Option 4 is correct.** When the load is doubled, the losses are quadrupled.

$$S_D = 200\,\text{MW} \quad P_{\text{LOSS}} = 1\,\text{MW}, \quad P_{\text{LOSS}} \propto (I)^2$$

$$S_D = 400\,\text{MW} \Rightarrow P_{\text{LOSS}} = 4\,\text{MW}$$

So the loss of the whole network (loss of two lines) is 8 MW.

1.16. **Option 2 is correct.** Because $\delta_2 < \delta_1$, then E_1 is generator and E_2 is motor.

1.17. **Option 2 is correct.** With the help of the power relationship and a KVL, we have

$$P_{12} = P_{d2} - P_{g2} = 2 - 1 = 1 = \frac{1 \times 1}{\frac{\sqrt{3}}{2}} \sin\delta \Rightarrow \delta = 60°$$

$$\hat{I} = \frac{\hat{V}_1 - \hat{V}_2}{j\frac{\sqrt{3}}{2}} = \frac{1\angle 60 - 1\angle 0}{j\frac{\sqrt{3}}{2}} = \frac{1\angle 120}{j\frac{\sqrt{3}}{2}} = \frac{1\angle 120}{\frac{\sqrt{3}}{2}\angle 90} = \frac{2}{\sqrt{3}}\angle 30$$

1.18. **Option 3 is correct.** Based on the answer to question 1.17 and the power relation:

$$\text{Production: } Q_2 = \frac{V_2^2 - V_1 V_2 \cdot \cos\delta}{X} = \frac{1 - 1 \times 1 \times \frac{1}{2}}{\frac{\sqrt{3}}{2}} = \frac{\frac{1}{2}}{\frac{\sqrt{3}}{2}} = \frac{1}{\sqrt{3}}$$

$$Q_2 = Q_{g2} - \sqrt{3} \Rightarrow Q_{g2} = \frac{1}{\sqrt{3}} + \sqrt{3} = \frac{4}{\sqrt{3}}$$

1.19. Option 2 is correct.

$$\max \delta = 60° \Rightarrow P_{\max}|_{\delta=60} = \frac{1 \times 1}{\dfrac{\sqrt{3}}{2}} \times \frac{\sqrt{3}}{2} = 1$$

1.20. Option 2 is correct.

$$S_b = 100\,\text{MVA}, \; V_{bg1} = 10\,\text{kV} \Rightarrow Z_b = \frac{V_b^2}{S_b} = \frac{(10\,\text{kV})^2}{100\,\text{MVA}} = 1\Omega \Rightarrow X_L\text{p.u.} = \frac{X_L}{Z_b} = \frac{1}{1} = 1$$

1.21. Option 2 is correct. From equations (1.25) and (1.26), we have:

$$Q_{C3ph} = P_{3ph}\left(\tan\left(\cos^{-1} pf_1\right) - \tan\left(\cos^{-1}(pf_2)\right)\right), \quad pf_1 = 0.6, \quad pf_2 = 0.8$$

From equation (1.59): $Q_{C3ph} = 240\,\text{MW}\left(\dfrac{4}{3} - \dfrac{3}{4}\right) = 140\,\text{MVAr}$

$$X_C = \frac{V_L^2}{Q_{C3ph}} \Rightarrow X_C = \frac{(10\,\text{kV})^2}{140\,\text{MVAr}} = \frac{5\Omega}{7}$$

1.22. Option 2 is correct.

$$S_n = \sqrt{3}V_n I_n \Rightarrow I_n = \frac{90\,\text{kVA}}{2\,K\sqrt{3} \times \sqrt{3}} = 15\,\text{A}$$

$$P = \sqrt{3}V_L I_L \cos\varphi \Rightarrow I = \frac{48\,\text{kW}}{\sqrt{3} \times 2K\sqrt{3} \times 0.8} = 10\,\text{A}$$

Voltage drop on the resistor: $= RI_n = \dfrac{1}{100} \times \dfrac{1}{\sqrt{3}} 2K\sqrt{3} \Rightarrow 15R = 20 \Rightarrow R = \dfrac{4}{3}$

Voltage drop on the reactance: $= XI_n = \dfrac{1}{100} \times \sqrt{3} \times 2K\sqrt{3} \Rightarrow 15X = 60 \Rightarrow X = 4$

$$P_{\text{line}} = 3RI^2 = 3 \times \frac{4}{3} \times 10^2 = 400\,\text{W}$$

$$Q_{\text{line}} = 3XI^2 = 3 \times 4 \times 10^2 = 1200\,\text{VAr}$$

1.23. Option 4 is correct.

We will consider : $S_b = 100\,\text{MVA}, \ V_{bG} = 20\,\text{kV}$

T_1 transformer secondary side : $V_{b1} = 20 \times \dfrac{140}{14} = 200\,\text{kV}$

Line: $Z_b = \dfrac{200\,\text{kV} \times 200\,\text{kV}}{100\,\text{MVA}} = 400\,\Omega \Rightarrow Z(\text{p.u.}) = \dfrac{20 + 100j}{400} = 0.05 + j0.25\,\text{p.u.}$

$Y_2(\text{p.u.}) = (0.2 + j2) \times 400 = 80 + j800$

1.24. Option 4 is correct. From equations (1.42), (1.43) and (1.44):

$\delta_2 > \delta_1 \rightarrow E_2$ is generator.

$E_1 = E_2 \rightarrow E_1$ and E_2 produce reactive power.

1.25. Option 3 is correct. From equation (1.29):

$Z_b = \dfrac{(20\,\text{kV})^2}{100\,\text{MVA}} = 4\,\Omega, \ Z_L = \dfrac{(20\,\text{kV})^2}{200\,\text{MVA}} = 2\,\Omega \Rightarrow Z(\text{p.u.}) = \dfrac{2\,\Omega}{4\,\Omega} = 0.5\,\text{p.u.}$

1.26. Option 1 is correct.

$P_{12} = 1 = \dfrac{1 \times 1}{0.1}\sin\delta \Rightarrow \sin\delta = 0.1$

Production: $Q_C - 1 = \dfrac{V_2^2 - V_1 V_2 \cdot \cos\delta}{0.1} = \dfrac{1 - \cos\left(\sin^{-1} 0.1\right)}{0.1}$

$= \dfrac{1 - 0.995}{0.1} = \dfrac{0.005}{0.1} = 0.05 \Rightarrow Q_C = 1 + 0.05 = 1.05$

1.27. Option 1 is correct.

$Q = P \cdot \tan\varphi \Rightarrow S_1 = 2 + j2\tan\left(\cos^{-1} 0.8\right), \ S_2 = 2 - j2\tan\left(\cos^{-1} 0.8\right), \ S_3 = 2$

$\Rightarrow S = S_1 + S_2 + S_3 = 6 \Rightarrow Y = \dfrac{S^*}{V^2} = \dfrac{6}{1} = 6$

1.28. **Option 2 is correct.**

$$X_{CA} = \frac{(V_L)^2}{\dfrac{Q_C}{3}} = \frac{(400\,\text{V})^2}{200\,\text{kVAr}} = 0.8 \Rightarrow C_\Delta = \frac{1}{2\pi f X_C} = \frac{1}{100\pi \times 0.8} \approx \frac{1}{100 \times 2.5}$$

$$= 4 \times 10^{-3} = 4000 \times 10^{-6} = 4000\,\mu\text{F}$$

1.29. **Option 3 is correct.** Consider a lagging load and its phasor diagram. If the delta angle increases, then the current angle (theta) must decrease. As a result, the voltage drop (XI) must increase, so the current increases.

$$\delta_2 > \delta_1 \Rightarrow \theta_2 < \theta_1 \Rightarrow XI_2 > XI_1 \Rightarrow I_2 > I_1$$

1.30. **Option 4 is correct.**

$$P = \frac{V_1 V_2}{X} \sin\delta = \frac{1 \times 1}{0.5} 0.25 = \frac{1}{2}$$

$$\text{Production:} \quad Q = \frac{V_2^2 - V_1 V_2 \cdot \cos\delta}{X} = \frac{1 - 0.96}{0.5} = \frac{0.04}{0.5} = 0.08$$

$$\text{Consumption:} \ Q = -0.08$$

1.31. **Option 2 is correct.**

$$V_{bload} = 20\,\text{kV} \times \left(\frac{200}{20}\right) \times \left(\frac{9}{180}\right)$$

$$V_{bload} = 10\,\text{kV}, \ \ S_b = S_G = 300\,\text{MVA}, \ \ Z_b = \frac{(10\,\text{kV})^2}{300\,\text{M}} = \frac{1}{3}\Omega$$

$$Z_{load} = \frac{(9\,\text{kV})^2}{180\,\text{MVA}} = 0.45\,\Omega, \ \ Z(\text{p.u.}) = \frac{0.45}{\dfrac{1}{3}} = 1.35$$

1.32. **Option 2 is correct.** From equation (1.11):

$$\frac{\Delta P}{P} = -2\sin\left(\cos^{-1}(0.8)\right)^2 \times \frac{10}{100} = -2 \times (0.6)^2 \times \frac{10}{100} = -0.072 = -7.2\%$$

1.33. **Option 4 is correct**. From equation (1.13):

$$\text{Impedance angle} = \cos^{-1}(pf) = \frac{R}{|Z|}$$

$$\text{If}: \quad |Z_1| = 1 \Rightarrow Z_1 = 0.8 + j0.6$$

If the frequency doubles we have:

$$Z_2 = 0.8 + j2 \times 0.6 = 0.8 + j1.2 \Rightarrow |Z_2| \approx \sqrt{2}$$

On the other hand, the voltage is constant.

$$|S_1| = \frac{V_1^2}{|Z_1|}, |S_2| = \frac{V_2^2}{|Z_2|} \Rightarrow \frac{|S_1|}{|S_2|} = \frac{|Z_2|}{|Z_1|} = \frac{\sqrt{2}}{1} \Rightarrow |S_2| = \frac{1}{\sqrt{2}}|S_1|$$

1.34. **Option 2 is correct**. From equation (1.12):

$$\frac{\Delta P}{P} = 2\frac{\Delta V}{V} \Rightarrow \frac{\Delta P}{P} = 2 \times 2 \Rightarrow \frac{\Delta P}{P} = 4$$

1.35. **Option 3 is correct**. From equation (1.45):

$$\delta = \tan^{-1}\frac{X}{R} \Rightarrow \delta = \tan^{-1}1 = 45°$$

1.36. **Option 1 is correct**.

$$V(t) = V_m \cos(\omega t + \theta_v), \quad i(t) = I_m \cos(\omega t + \theta_i)$$

$$P(t) = \frac{1}{2}V_m I_m \cos\varphi + \frac{1}{2}V_m I_m \cos(2\omega t + \theta_v + \theta_i)$$

$$\varphi = \theta_v - \theta_i$$

We have:

$$\cos\varphi = \frac{800}{1000} = 0.8 \Rightarrow \varphi = \pm36°, \quad \phi_0 > 0 \Rightarrow \left\{ \begin{array}{c} \theta_v + \theta_i = -\phi_0 \\ \theta_v = 0 \end{array} \right\} \Rightarrow \theta_i = -36°$$

Then $Q > 0$ and:

$$S = 800 + j1000 \times \sin(36) = 800 + j600$$

1.37. Option 3 is correct.

$$R = \frac{V^2}{P} = \frac{200 \times 200}{1.2\,\text{kW}} = 33.33\,\Omega \Rightarrow Q = P\tan\phi = 1.2k\tan\left(\cos^{-1}0.6\right)$$

$$= 1.2\,k \times \tan(53) = 1.6\,\text{kV} \Rightarrow X = \frac{V^2}{Q} = \frac{(200)^2}{1.6\,\text{kVAr}} = 25\,\Omega$$

1.38. Option 4 is correct.

$$PF_Y = PF_\Delta \Rightarrow Q_Y = Q_\Delta \Rightarrow 3 \times \frac{\left(\dfrac{V_L}{\sqrt{3}}\right)^2}{X_{CY}} = 3 \times \frac{V_L^2}{X_{C\Delta}} \Rightarrow 3X_{CY} = X_{C\Delta}$$

1.39. Option 2 is correct.

$$Q_2 = \frac{V_2^2 - V_1 V_2 \cdot \cos\delta}{X}, \quad P = \frac{V^2}{R} = \frac{(10\,\text{kV})^2}{1} = 100\,\text{MW} = \frac{V_1 V_2}{X}\sin\delta$$

$$100\,\text{MW} = \frac{(10\,\text{kV})^2}{1}\sin\delta \Rightarrow \sin\delta = 1 \Rightarrow \delta = 90$$

$$\text{Production: } Q = \frac{(10\,\text{kV})^2 - (10\,\text{kV})^2\cos 90^\circ}{1} = 100\,\text{MVAr}$$

$$X_C = \frac{V^2}{Q} = \frac{(10\,\text{kV})^2}{100\,\text{MVAr}} = 1$$

$$\Rightarrow C = \frac{1}{2\pi f X_C} = \frac{1}{100\pi \times 1} = \frac{1}{300} = 33 \times 10^{-4} = 3300\,\mu\text{F}$$

1.40. Option 2 is correct. We have two scenarios

Scenario 1: $\theta_i > 0 \Rightarrow \hat{I} = a + jb \Rightarrow V_2 < \theta_v = V_1 < 0 - jX(a + jb)$

Scenario 2: $\theta_i < 0 \Rightarrow \hat{I} = a - jb \Rightarrow V_2 < \theta_v = V_1 < 0 - jX(a - jb)$

Scenario 1: $V_2 < \theta_v = (V_1 + Xb) - jXa \Rightarrow V_2 = \sqrt{(V_1 + Xb)^2 + (Xa)^2}$

Scenario 2: $V_2 < \theta_v = (V_1 Xb) - jXa \Rightarrow V_2 = \sqrt{(V_1 Xb)^2 + (Xa)^2}$

In the first scenario, it is always $(V1 < V2)$.

1.41. **Option 3 is correct.**

At lagging load, $V1$ always becomes larger than $V2$.

1.42. **Option 3 is correct.**

$$P_2 = P_{21} = P_{g2} - P_{d2} = 1 - 2 = -1 \Rightarrow P_{12} = 1$$

$$P_{12} = 1 = \frac{1 \times V_2}{0.5} \sin \delta \Rightarrow V_2 \sin \delta = 0.5$$

$$Q_2 = \frac{V_2^2 - V_1 V_2 \cdot \cos \delta}{X} = 0 \Rightarrow V_2 = \cos \delta$$

$$\sin \delta \cos \delta = 0.5 \Rightarrow \sin 2\delta = 1 \Rightarrow \delta = 45° \Rightarrow V_2 = \frac{\sqrt{2}}{2} = \sqrt{0.5}$$

1.43. **Option 2 is correct.** From equation (1.39):

$$Q_{g1} - Q_{d1} = Q_1 = \frac{V_1^2 - V_1 V_2 \times \cos \delta}{X} = \frac{1 - 0.707 \times 1 \times 0.707}{0.5} = 1 \text{ p.u.}$$

$$Q_{g1} - 1 = 1 \text{ p.u.} \Rightarrow Q_{g1} = 2$$

We have the figure: $Q_2 = Q_{g2} + Q_{d2} = 0 + 0 = 0$
From equation (1.40):

$$Q_{\text{Line}} = Q_1 + Q_2 = 1 + 0 = XI^2 = 0.5I^2 \Rightarrow I^2 = 2 \Rightarrow I = \sqrt{2}$$

1.44. **Option 2 is correct.**

$$Q = \frac{V_2 V_1 \cdot \cos \delta - V_2^2}{X} > 0 \Rightarrow (V_1 \cos \delta - V_2) > 0$$

$$\Rightarrow \cos \delta > \frac{V_2}{V_1} \Rightarrow \cos \delta > \frac{1}{2} \Rightarrow \delta < 60$$

1.45. **Option 4 is correct.**

In this example, $I1$ is the current of the generator, $I2$ is the current of load 2, and $I3$ is the current of load 3:

$$\hat{I}_3 = \frac{2}{1} = 2\,\text{p.u.} \Rightarrow \hat{V}_2 = j0.5 \times 2 + 1 = 1 + j \Rightarrow \hat{I}_2 = \frac{S_2^*}{\hat{V}_2^*} = \frac{1-j}{1-j} = 1$$

$$\hat{I}_1 = 1 + 2 = 3\,\text{p.u.} \Rightarrow \hat{V}_1 = \hat{V}_2 + j0.5 \times \hat{I}_1 = 1 + j + 1.5j = 1 + j2.5$$

1.46. **Option 4 is correct.**

According to Scenario 1, the line does not affect bus 2 admittance and then:

$$Y_{22} = \left.\frac{\hat{I}_2}{\hat{V}_2}\right|_{I_1=0} = -j2 \Rightarrow Z_{22} = j0.5 = Z_{\text{Load}}$$

Current to bus1 voltage ratio in Scenario 2 includes both line and load admittance.

$$Y_{11} = \left.\frac{\hat{I}_1}{\hat{V}_1}\right|_{I_2=0} = -j1 \Rightarrow Z_{11} = j1 = jX + Z_{\text{load}} \Rightarrow X = 0.5$$

1.47. **Option 4 is correct.** From equation (1.43) and $|V_1| = |V_2| = |V_3|$ and

$$Q_{13} = Q_{31} = \frac{Q_{\text{Line 13}}}{2}$$

1.19 Two-Choice Questions (Yes – No) – 106 Questions

1. With an AC system, you can easily generate more power at a higher voltage than with a DC system.

2. Ultra-high-voltage AC is more economical to transfer than ultra-high voltage DC over very long distances.

3. With an inductor and capacitor, the single-phase resistor can be balanced once.

4. DC communication is characterized by the production of harmonics, which must be filtered, as well as compensating for a lot of reactive power on both sides of the line.

5. Substations are usually equipped with capacitor banks and reactors to maintain line voltage.

6. Industrial loads are independent of frequency and are a function of voltage.

7. Commercial and residential loads are highly dependent on frequency.

8. The load factor is the ratio of average load over a given period of time to peak load.

9. The load factor of a power plant must be low in order for it to operate economically.

10. The frequency of the instantaneous power is twice the frequency of the source if the voltage and current are sine waves.

11. In passive networks, negative power is the absorption of energy by inductors or capacitors.

12. The current is ahead of the voltage when the load is inductive.

13. A power's apparent magnitude corresponds to a voltage's effective amount multiplied by a current's effective amount.

14. Power factor is leading when the current is behind the voltage.

15. Both P and Q have the same unit.

16. In the case of a pure capacitor, the energy is converted from electrical to non-electric.

17. Complex power is apparent power.

18. Apparent power directly indicates heating.

19. The apparent power angle for capacitive loads is negative.

20. We always have effective values: $\dfrac{|V|^2}{\hat{S}} = Z$

21. The power company seeks power coefficients near zero for the main loads of its system.

22. Active power distribution is not affected by the voltage angle difference between the terminals.

23 At the same load, production power equals negative power consumption.

24. In a three-phase system, the instantaneous power delivered to the load is constant.

25. In the power system, the generators have a triangular connection.

26. If phase B is ahead of phase A, the phase sequence is negative.

27. The following voltage sequence is a negative sequence.

$$\hat{V}_{an} = V\angle 0, \quad \hat{V}_{bn} = V\angle -120, \quad \hat{V}_{cn} = V\angle -240$$

28. A phase sequence is negative if V_{ab} exceeds V_a by 30°.

29. This relationship is correct: $Z_Y = 3Z_\Delta$

30. Neutral wire resistance change has no effect on the balanced system.
31. In the opposite relation, V_{Peak} is the line voltage.

$$P_{3ph} = 1.5\, V_{Peak}\, I_{Peak}\, \cos\varphi$$

32. Whether you are in triangle mode or star mode has no effect on your power.
33. Being a star or a triangle of constant impedance load does not affect the calculations.
34. It doesn't matter whether you are a star or a triangle with a constant power load.
35. The transformer can increase or decrease the voltage of the buses by adjusting the tap changer.
36. To fully define a per-unit system, at least five base quantities must be defined.
37. We have: $P_{3ph}(\text{p.u.}) = \sqrt{3}V(\text{p.u.})I(\text{p.u.})\cos\varphi$.
38. We have: $Z(\text{p.u.}) = \left|\dfrac{V_{l-l}}{V_{base}}\right|^2 \times \dfrac{S_{base}\angle\varphi}{S_{3ph}^*}$.
39. If the base voltages are the same, we have: $Z_{\text{p.u.}}^{new} = Z_{\text{p.u.}}^{old} \times \dfrac{S_{base}^{new}}{S_{base}^{old}}$.
40. In a per-unit system, the primary and secondary currents of the transformer are equal.
41. The per-unit system maintains the unit of parameters.
42 The base voltage and power on the transformer side are equal.
43. At an angle of 45°, maximum transmission power occurs in a single-feed power system.
44. The parallel capacitor on the transmission line corrects the power factor and increases the voltage.
45. When the load is either inductive or capacitive, the instantaneous power of the positive and negative half-cycles is the same, and their mean is zero.
46. Current delay in relation to voltage is the leading power factor.
47. Power factor is: $\cos\varphi = \dfrac{P}{\sqrt{P^2 + Q^2}}$
48. Capacitors deliver lagging current or receive leading current.
49. Reactive power consumption is positive in capacitive loads.
50. In a balanced three-phase circuit: $V_{l-l} = \sqrt{3} \times V_{ph}$.

51. The total power of a three-phase generator or the total power consumed by a three-phase load equals the total power of all three phases.

52. In a three-phase system: $\left|S_{3\,\text{ph}}\right| = \sqrt{3}V_{l-l}I_{l}$.

53. $V_{l-l}(\text{p.u.}) = V_{\text{ph}}(\text{p.u.})$

54. In the load flow study, the location of relays and circuit breakers does not matter.

55. When an asymmetric short circuit occurs, knowing the points of connection to ground is not necessary to calculate the ground current.

56. Active and reactive power in a transmission network is almost independent of one another.

57. The control of active power depends on voltage control and the control of reactive power depends on frequency control.

58. The frequency of a system depends on its active power balance.

59. On a per-unit system, balanced generator speed is a display of system frequency.

60. Reactive power can be absorbed or fed into overhead transmission lines through long, lossless lines dependent on the load current.

61. In loads less than natural load (surge impedance loading), reactive power lines absorb.

62. In loads greater than natural load (surge impedance loading), reactive power lines produce.

63. Due to their high capacitance, underground cables have high natural loads.

64. Underground cables are always loaded above their normal capacity.

65. Reactive power is always absorbed by transformers regardless of load.

66. A high "leading power factor" causes excessive voltage drops in the transmission network.

67. At all levels of the system, voltage level control is achieved by controlling the generation and absorption of reactive power current.

68. The main factor in controlling voltage is the generator.

69. The system voltage is controlled by series capacitors.

70. Synchronous condensers are active compensation devices for mains voltage control.

71. Shunt reactors compensate for the effects of line inductors.

72. Short circuits are limited by shunt reactors.

73. In high-voltage overhead lines longer than 200 km, shunt reactors are required.

74. If the voltage and frequency drop at the same time at the impedance load ($R = X$), the power will also decrease.

75. As a result of the Ferranti effect, the receiver side voltage is reduced.

76. Shunt capacitors increase local voltages and provide reactive power.

77. The main disadvantage of shunt capacitors is that they produce less reactive power at low voltages.

78. By correcting the power factor, reactive power is provided close to consumption instead of coming from distant sources.

79. Transmission systems use shunt capacitors to compensate for line losses.

80. Under high load conditions, shunt capacitors ensure that voltage levels remain stable.

81. For compensation of line reactance, series capacitors are used.

82. With increasing power transfer, a series capacitor's reactive power increases.

83. The goal of transmission line compensation is complete compensation.

84. Voltage index and the effect on power transfer capability are critical factors in choosing the location of the series capacitor.

85. Synchronous condensers rotate without mechanical load.

86. In transformers, synchronous condensers are often connected to the third coil.

87. Shunt capacitor compensation is generally the most economical source of reactive power.

88. Combining series and shunt capacitors allows independent control of characteristic impedance and load angle.

89. Passive compensating devices are modeled as fixed admittance elements in load flow studies.

90. Active power can be seen in the relation to average power.

91. Reactive power can be seen in the relation to average power.

92. For a single-phase system: $S = \hat{V}\hat{I}$

93. Positive reactive power is produced by the leading load.

94. Currents b and c (+120) have an angle difference with current a in a three-phase system.

95. In single-phase per-unit system: $Z_b V_b^2 = S_b$

96. If the base voltage and power are doubled in a per-unit system, the per-unit impedance is halved.

97. It is possible to determine the active power of an asymmetric three-phase system without a neutral wire by using two wattmeter.

98. As far as the two AC machines are concerned, the generator is the one with the highest voltage.

99. Per-unit systems are assumed to have a base power equal to their largest rated power.

100. A 1% drop in voltage increases the power consumption by 2% at impedance load.

101. When the angle between the two feeds is zero (delta=zero), the load's reactive power is zero.

102. In a dual feed system, maximum active power is transmitted at a 90° angle if $R=X$.

103. It is possible for the transmission power to be reduced if one of the parallel lines is out.

104. If the frequency of a resistance load is doubled, the power consumption is also doubled.

105. The thermal limit is more important than the static limit in the short line.

106. Static limits are less important in long lines than dynamic limits.

1.20 Key Answers to Two-Choice Questions

1. Yes
2. No
3. Yes
4. Yes
5. Yes
6. No
7. No
8. Yes
9. No
10. Yes
11. Yes
12. No
13. Yes
14. No
15. Yes

16. No
17. No
18. No
19. No
20. No
21. No
22. No
23. Yes
24. Yes
25. No
26. Yes
27. No
28. No
29. No
30. Yes
31. No
32. No
33. No
34. Yes
35. Yes
36. No
37. No
38. No
39. Yes
40. Yes
41. No
42. No
43. Yes
44. Yes
45. Yes
46. Yes
47. Yes
48. Yes
49. No
50. Yes
51. No

52. Yes
53. Yes
54. Yes
55. No
56. Yes
57. No
58. Yes
59. Yes
60. Yes
61. No
62. No
63. Yes
64. No
65. Yes
66. No
67. Yes
68. Yes
69. No
70. Yes
71. No
72. No
73. Yes
74. Yes
75. No
76. Yes
77. Yes
78. Yes
79. Yes
80. Yes
81. Yes
82. Yes
83. No
84. Yes
85. Yes
86. Yes
87. Yes

88. Yes
89. Yes
90. Yes
91. No
92. No
93. Yes
94. No
95. No
96. Yes
97. Yes
98. No
99. Yes
100. No
101. No
102. No
103. Yes
104. No
105. Yes
106. Yes

2

Transmission Line Parameters

Part One: Lesson Summary

2.1 Introduction

From the perspective of power system analysis, series inductance and parallel capacitance are two of the most critical design parameters of a transmission line. This chapter describes methods for calculating these parameters for single-phase and three-phase systems. Throughout this chapter, we have discussed the line as a component of a symmetrical power system, which can be analyzed using the single-phase method.

This chapter introduces the concepts of inductance and capacitance, solid round conductor, hollow thin-walled conductor, stranded conductor, bundle, three-phase transmission transposed line, geometric mean distance (GMD), geometric mean radius (GMR), earth effect, double-circuit three-phase, Corona effect, three-phase transmission line, reactive compensation, short, medium, long-length transmission line, and so on.

This chapter, as well as the previous and subsequent chapters, contains problems that can be solved mentally without using a calculator.

2.2 Line Resistance

$$R_{dc} = \frac{\rho \ell}{A} \qquad (2.1)$$

ρ: Conductor resistivity
ℓ: Conductor length
A: Conductor cross-sectional area

Effect of temperature on resistance

$$R_2 = R_1 \frac{T + t_2}{T + t_1} \tag{2.2}$$

T: Temperature constant (for aluminum $T=228$)
R_1: Conductor resistance at t_1 ($^\circ$C)
R_2: Conductor resistance at t_2 ($^\circ$C)

Skin effect:

$$R_{AC} = 1.02 \times R_{DC}$$

2.3 Line Inductance

Internal inductance

$$\mu_0 = 4\pi \times 10^{-7} \tag{2.3}$$

$$L_{int} = \frac{\mu_0}{8\pi} = \frac{1}{2} \times 10^{-7} \, (\text{H/m}) \tag{2.4}$$

External inductance, inductance between two points external

$$L_{ext} = 2 \times 10^{-7} \ln \frac{D_2}{D_1} \tag{2.5}$$

The flux linkage, self, mutual inductance, and voltage of conductor (i) in a group of n conductors

$$V_i = \lambda_i \, \omega, \quad \lambda_i = L_{ii} \hat{I}_i + \sum_{j=1, j \neq i}^{n} \left(L_{ij} \hat{I}_j \right) \tag{2.6}$$

$$\lambda_i = 2 \times 10^{-7} \left(\hat{I}_i \ln \frac{1}{D_{ii}} + \sum_{j=1, j \neq i}^{n} \hat{I}_j \ln \frac{1}{D_{ij}} \right) \tag{2.7}$$

where D_{ii} is the geometric mean radius (GMR) and:

$$D_{ii} = \begin{cases} r: & \text{hollow thin walled conductor} \\ r' = r \cdot e^{-0.25} = 0.7788r: & \text{solid round conductor} \\ D_s: & \text{stranded conductor} \end{cases} \tag{2.8}$$

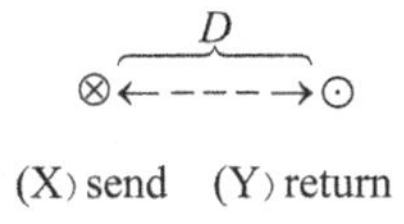

$$\overset{D}{\underset{(X)\ \text{send} \quad (Y)\ \text{return}}{\otimes \longleftarrow ---- \longrightarrow \odot}}$$

FIGURE 2.1
Single-phase line.

2.3.1 Single-Phase Line Inductance

The inductance of conductor X (send) (Figure 2.1)

$$L_x = 2 \times 10^{-7} \ln \frac{D}{D_{sx}} (\text{H/m})$$

$$= 0.2 \ln \frac{D}{D_{sx}} (\text{mH/km}) \tag{2.9}$$

The inductance of conductor Y (return)

$$L_y = 2 \times 10^{-7} Ln \frac{D}{D_{sy}} (\text{H/m})$$

$$= 0.2 Ln \frac{D}{D_{sy}} (\text{mH/km}) \tag{2.10}$$

where D_{sx} and D_{sy} can be obtained using equation (2.8).
    The total inductance of a single-phase line

$$L_{tot} = L_x + L_y \tag{2.11}$$

2.3.2 The Total Inductance of a Single-Phase Consisting of Two Composite Conductors

(X) the group consists of n conductors and (Y) group consists of m conductors (Figure 2.2).

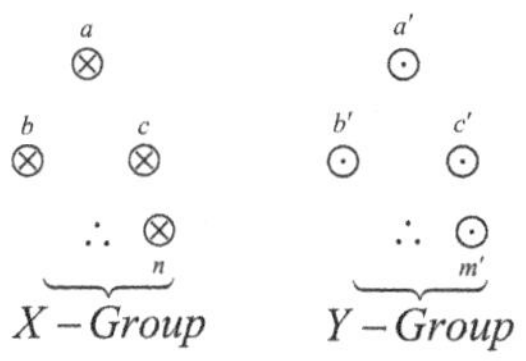

FIGURE 2.2
Group single-phase line.

$$L_x = 2 \times 10^{-7} \ln \frac{\text{GMD}}{\text{GMR}_x}$$
$$L_y = 2 \times 10^{-7} \ln \frac{\text{GMD}}{\text{GMR}_y}$$
$$\left. \right\} \Rightarrow L_{\text{tot}} = L_x + L_y \qquad (2.12)$$

where:

$$\text{GMD} = \sqrt[m \times n]{\left(D_{aa'} \cdot D_{ab'} \cdots D_{am'}\right) \cdots \left(D_{na'} \cdot D_{nb'} \cdots D_{nm'}\right)} \qquad (2.13)$$

$$\text{GMR}_x = \sqrt[n^2]{\left(D_{aa} \cdot D_{ab} \cdots D_{an}\right) \cdots \left(D_{na} \cdot D_{nb} \cdots D_{nn}\right)} \qquad (2.14)$$

$$\text{GMR}_y = \sqrt[m^2]{\left(D_{a'a'} \cdot D_{a'b'} \cdots D_{a'm'}\right) \cdots \left(D_{m'a'} \cdot D_{m'b'} \cdots D_{m'm'}\right)} \qquad (2.15)$$

$$D_{aa} = D_{bb} = \cdots = D_{nn} = D_{sx} \qquad (2.16)$$

$$D_{a'a'} = D_{b'b'} = \cdots = D_{m'm'} = D_{sy} \qquad (2.17)$$

Note:
In the X-group, the GMR_x is the n^2th root of the product of n internal distances between n conductors.

In the Y-group, the GMR_y is the m^2th root of the product of m internal distances between m conductors.

The GMD is the $m \times n$th root of the product of n distances between n conductors in the X-group and m conductors in the Y-group.

$$\text{GMD} = \sqrt[m \times n]{\underbrace{\underbrace{\left(D_{aa'} \cdot D_{ab'} \ldots D_{am'}\right)}_{m} \ldots \underbrace{\left(D_{na'} \cdot D_{nb'} \ldots D_{nm'}\right)}_{m}}_{n}} \qquad (2.18)$$

2.3.3 The Inductance of Three-Phase, Untransposed Transmission Line ($L_a \neq L_b \neq L_c$)

$$L_a = 2 \times 10^{-7} \left(\ln \frac{1}{D_s} + \alpha^2 \ln \frac{1}{D_{12}} + \alpha \ln \frac{1}{D_{13}} \right)$$

$$L_b = 2 \times 10^{-7} \left(\alpha \ln \frac{1}{D_{12}} + \ln \frac{1}{D_s} + \alpha^2 \ln \frac{1}{D_{23}} \right) \qquad (2.19)$$

$$L_c = 2 \times 10^{-7} \left(\alpha^2 \ln \frac{1}{D_{13}} + \alpha \ln \frac{1}{D_{23}} + \ln \frac{1}{D_s} \right)$$

where (Figure 2.3):

$$\alpha^2 = 1\angle 240°, \quad \alpha = 1\angle 120° \qquad (2.20)$$

FIGURE 2.3
Three-phase untransposed transmission line.

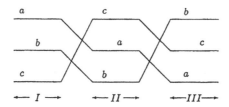

FIGURE 2.4
Three-phase transposed transmission line.

2.3.4 The Inductance of Three-Phase, Transposed Transmission Line ($L_a = L_b = L_c$)

The average inductance of each phase is equal to (Figure 2.4):

$$L_{av} = \frac{L_a + L_b + L_c}{3} = L \tag{2.21}$$

The inductance of each phase is:

$$L = 0.2 \ln \frac{GMD}{D_s} (mH/km) = L_a = L_b = L_c \tag{2.22}$$

$$GMD = \sqrt[3]{D_{12} \cdot D_{13} \cdot D_{23}} \tag{2.23}$$

2.3.5 GMR of Bundle Conductors

The electrical connection of sub-conductors in a group, such as the following conductors, is referred to as a "bundle conductor" (Figure 2.5).

FIGURE 2.5
Two, three, and four sub-conductor bundles.

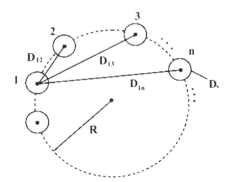

FIGURE 2.6
(n) sub-conductor bundle - $\mathrm{GMR}_n = \sqrt[n]{n \cdot D_s \cdot R^{n-1}}$.

$$\mathrm{GMR}_2 = \sqrt{D_s \cdot d}, \quad \mathrm{GMR}_3 = \sqrt[3]{D_s \cdot d^2}, \quad \mathrm{GMR}_4 = \sqrt[4]{D_s \cdot d^3 \cdot \sqrt{2}}$$

where d is the bundle spacing and D_s is the GMR of each sub-conductor
(Figure 2.6).

2.3.6 Inductance of Bundle Conductors

GMD is calculated without including the effect of the bundle. When the
distance between conductors in each phase is much less than the distance
between phases, GMD=D.

2.3.7 Inductance of Three-Phase Double-Circuits, Transposed Lines

$$L = 0.2 \ln \frac{\mathrm{GMD}}{\mathrm{GMR}} (\mathrm{mH/km}) \tag{2.24}$$

$$\mathrm{GMD} = \sqrt[3]{D_{AB} D_{BC} D_{AC}} \tag{2.25}$$

where (Figure 2.7):

$$D_{AB} = \sqrt[4]{D_{a_1 b_1} D_{a_1 b_2} D_{a_2 b_1} D_{a_2 b_2}} \tag{2.26}$$

$$D_{AC} = \sqrt[4]{D_{a_1 c_1} D_{a_1 c_2} D_{a_2 c_1} D_{a_2 c_2}} \tag{2.27}$$

$$D_{BC} = \sqrt[4]{D_{b_1 c_1} D_{b_1 c_2} D_{b_2 c_1} D_{b_2 c_2}} \tag{2.28}$$

$$\mathrm{GMR} = \sqrt[3]{D_{SA} D_{SB} D_{SC}} \tag{2.29}$$

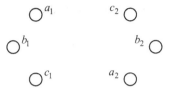

FIGURE 2.7
Three-phase double-circuits, transposed lines.

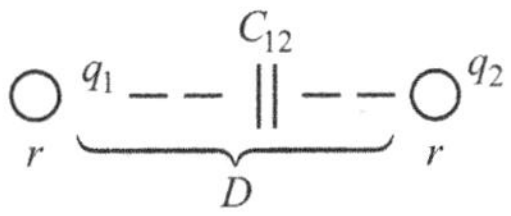

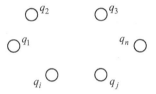

FIGURE 2.8
Two conductors in a single-phase line.

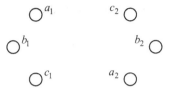

FIGURE 2.9
Analysis (n) of a conductor in space.

Where:

$$D_{SA} = \sqrt{D_s D_{a_1 a_2}}, \quad D_{SB} = \sqrt{D_s D_{b_1 b_2}}, \quad D_{SC} = \sqrt{D_s D_{c_1 c_2}} \tag{2.30}$$

If the conductors are bundled, D_S is calculated as Section 2.3.5 GMR of the bundled conductors.

2.4 Line Capacitance

$$C = \frac{q}{V} \tag{2.31}$$

2.4.1 The Voltage between Two Conductors in a Single-Phase Line (Figure 2.8)

$$V_{12} = \frac{q}{\pi \varepsilon_0} \ln \frac{D}{r}, \quad q_1 = -q_2 = q \tag{2.32}$$

2.4.2 The Capacitance between Two Conductors in a Single-Phase Line

$$C_{12} = \frac{\pi \varepsilon_0}{\ln \dfrac{D}{r}} (\text{F}/\text{m}) \tag{2.33}$$

2.4.3 Capacitance between a Conductor and a Neutral

A capacitor between two conductors is equivalent to two capacitors in series between each phase and neutral (2.35).

$$C = \frac{2\pi \varepsilon_0}{\ln \dfrac{D}{r}} (\text{F}/\text{ m}) \simeq \frac{0.0556}{\ln \dfrac{D}{r}} (\mu\text{F}/\text{km}) \simeq \frac{1}{18\ln \dfrac{D}{r}} (\mu\text{F}/\text{km}) \tag{2.34}$$

$$C = 2C_{12} \tag{2.35}$$

2.4.4 Potential Differences between Two Conductors (Figure 2.9)

$$V_{ij} = \frac{1}{2\pi \varepsilon_0} \sum_{k=1}^{n} q_k \ln \frac{D_{kj}}{D_{ki}}, \quad D_{ii} = r(\text{radius}) \tag{2.36}$$

2.4.5 Potential Difference between a Conductor and Earth

$$V_i = \frac{1}{2\pi \varepsilon_0} \sum_{k=1}^{n} q_k \ln \frac{1}{D_{ik}}, \quad D_{ii} = r \tag{2.37}$$

2.4.6 Capacitance of a Single-Phase Line with Group Conductors

Two capacitors (C_x) and (C_y) are defined if (x) is the send group conductors and (y) is the return group conductors.

$$C_{xy} = \frac{C_{xn} \cdot C_{yn}}{C_{xn} + C_{yn}} = \frac{\pi \cdot \varepsilon_0}{\ln \dfrac{\text{GMD}}{\sqrt{\text{GMR}_x \cdot \text{GMR}_y}}} \tag{2.38}$$

where:
   Capacitance of send conductors to the earth:

$$C_{xn} = \frac{2\pi \varepsilon_0}{\ln \dfrac{\text{GMD}}{\text{GMR}_x}} \tag{2.39}$$

Capacitance of return conductors to the earth:

$$C_{yn} = \frac{2\pi\varepsilon_0}{\ln\dfrac{\text{GMD}}{\text{GMR}_y}} \tag{2.40}$$

In which they can be obtained using equations (2.13) and (2.14) with the difference that whether conductors are solid, hollow or string we always have $D_{aa} = D_{bb} = D_{cc} = r$.

2.4.7 Capacitance of Three-Phase Transposed Lines

$$C = \frac{2\pi\varepsilon_0}{\ln\dfrac{\text{GMD}}{\text{GMR}}} \tag{2.41}$$

$$\text{GMD} = \sqrt[3]{D_{12} \cdot D_{23} \cdot D_{13}} \tag{2.42}$$

2.4.8 Capacitance of Three-Phase Double-Circuits, Transposed Lines

$$C = \frac{2\pi\varepsilon_0}{\ln\dfrac{\text{GMD}}{\text{GMR}}} \tag{2.43}$$

For calculating GMD and GMR, we use equations (2.25)–(2.30), with the difference that it is always $D_S = r$.

2.4.9 Effect of Earth on the Capacitance

By drawing lines relative to the earth, we can calculate the effect of the earth on the capacitance, causing the capacitance to increase.

2.4.10 Effect of Earth on the Capacitance in a Single-Phase Line

$$C_{an} = \frac{2\pi\varepsilon_0}{\ln\dfrac{D \times 2H_1}{r \cdot \sqrt{D^2 + 4H_1 \cdot H_2}}} \tag{2.44}$$

You can calculate C_{bn} by replacing H_1 with H_2 (Figure 2.10).

FIGURE 2.10
Effect of earth on the capacitance in a single-phase line.

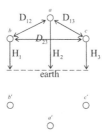

FIGURE 2.11
Effect of earth on the capacitance in a three-phase line.

2.4.11 Effect of Earth on the Capacitance in a Three-Phase Line

If a', b', and c' are images of phases a, b, and c on the earth, respectively, then:

$$C_{an} = \frac{2\pi\varepsilon_0}{\ln \dfrac{\text{GMD} \times H}{\text{GMR} \times H_S}} \qquad (2.45)$$

where (Figure 2.11):

$$H = 2 \times \sqrt[3]{H_1 \cdot H_2 \cdot H_3}$$

$$H_S = \sqrt[3]{D_{ab'} \cdot D_{ac'} \cdot D_{bc'}}$$

2.5 Corona Phenomenon

Whenever a voltage gradient (electrical field) is greater than the electrical field of the surrounding air, ionization occurs at the surface of a conductor. By bundling the line and increasing the number of conductors or the cross-section of the conductors, this phenomenon can be prevented to some extent.

2.6 Special Double-Circuits Lines

Assuming D_{Sb} is the GMR of the bundle, we have (Figures 2.12–2.15):

Figure 2.12:
$$L = 10^{-7} \ln\left(\frac{\sqrt{3}D}{2 \cdot D_{sb}}\right) (\text{H/m}) \tag{2.46}$$

Figure 2.12:
$$C = \frac{4\pi\varepsilon_0}{\ln\left(\dfrac{\sqrt{3}D}{2r}\right)} (\text{F/m}) = \frac{1}{9\ln\left(\dfrac{\sqrt{3}D}{2r}\right)} (\mu\text{F/km}) \tag{2.47}$$

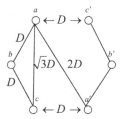

FIGURE 2.12
A regular hexagon, three-phase double circuit (Type 1).

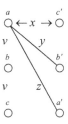

FIGURE 2.13
A vertical hexagon, three-phase double circuit (Type 1).

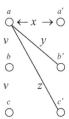

FIGURE 2.14
A vertical hexagon, three-phase double circuit (Type 2).

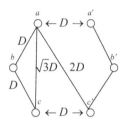

FIGURE 2.15
A regular hexagon, three-phase double circuit (Type 2).

Figures 2.13: $L = 2 \times 10^{-7} \ln\left(2^{\frac{1}{6}} \cdot \left(\frac{v}{D_{sb}}\right)^{\frac{1}{2}} \cdot \left(\frac{y}{z}\right)^{\frac{1}{3}} \right)$ (H/m)   (2.48)

Figures 2.13: $C = \dfrac{2\pi\varepsilon_0}{\ln\left(2^{\frac{1}{6}} \cdot \left(\frac{v}{r}\right)^{\frac{1}{2}} \cdot \left(\frac{y}{z}\right)^{\frac{1}{3}} \right)}$ (F/m)   (2.49)

Figures 2.14: $\dfrac{\text{GMD}}{\text{GMR}} = \dfrac{2^{\frac{1}{6}} \times v^{\frac{1}{2}} \times y^{\frac{1}{3}} \times z^{\frac{1}{6}}}{D_{sb}^{\frac{2}{}} \times x^{\frac{1}{2}}}$ (2.50)

Figures 2.15: $L = 0.1\ln\left(\dfrac{\sqrt{3}D}{D_{sb}} \right)$ (mH/km)   (2.51)

Part Two: Answer Question

2.7 Four-Choice Questions – 35 Questions

2.1. In the three-phase double-circuit line of Figure 2.16, which is the inductance of each phase in terms of (mH/km) (the GMR of each conductor is D_S)?

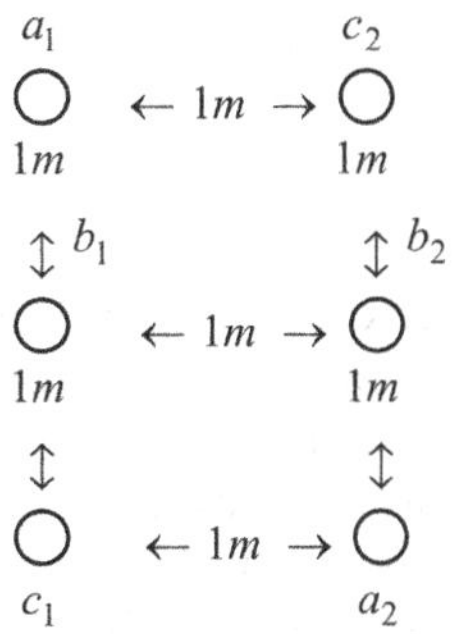

$$a_1 \quad\quad\quad\quad c_2$$
$$\bigcirc \; \leftarrow 1m \rightarrow \; \bigcirc$$
$$1m \quad\quad\quad\quad 1m$$

$$\updownarrow b_1 \quad\quad\quad \updownarrow b_2$$
$$\bigcirc \; \leftarrow 1m \rightarrow \; \bigcirc$$
$$1m \quad\quad\quad\quad 1m$$

$$\updownarrow \quad\quad\quad\quad \updownarrow$$
$$\bigcirc \; \leftarrow 1m \rightarrow \; \bigcirc$$
$$c_1 \quad\quad\quad\quad a_2$$

FIGURE 2.16
A vertical hexagon, three-phase double circuit.

1. $0.2\ln\dfrac{2^{\frac{5}{3}}}{D_S^{\frac{1}{2}}5^{\frac{1}{3}}}$

2. $0.2\ln\left(2^{\frac{1}{3}}D_S^{-\frac{1}{2}}5^{-\frac{1}{6}}\right)$

3. $0.2\ln\left(2^{\frac{1}{3}}D_S^{\frac{1}{2}}5^{\frac{1}{3}}\right)$

4. $0.2\ln\dfrac{2^{\frac{1}{3}}}{D_S^{\frac{1}{2}}5^{\frac{1}{3}}}$

2.2. In the conductors x, y, and z in Figure 2.17, three-phase currents are balanced. D_S is the GMR for each conductor and equals half the distance between each conductor. Single-phase transmission lines use conductors z and x to carry send currents, while conductor y carries return current, and its current is double the send current. In this case, what is the ratio of inductance between the three-phase arrangement and the single-phase arrangement?

FIGURE 2.17
A horizontal three-phase (Question network 2.2).

1. $\dfrac{5}{6}$ 2. $\dfrac{4}{3}$ 3. 2   4. $\dfrac{7}{4}$

2.3. A three-phase transmission line is on the vertices of an equilateral triangle to side D. Three-phase conductors with a triple bundle and with equal distance d are used. In this case, the radius of each string is r. If we want to make a transmission line from the above transmission line, whose capacitance (C) is equal to the mentioned line, and a double bundle is used, how much should the distance between the strings in each phase be? Do not consider the earth's effect.

1. $d\sqrt[3]{\dfrac{1}{r}}$ 2. $d\sqrt[3]{\dfrac{d}{r}}$ 3. $d\sqrt{\dfrac{1}{2}}$ 4. $r\sqrt[3]{\dfrac{r}{d}}$

2.4. According to Figure 2.18, the radius of a three-phase transmission line is $\sqrt[3]{2}$ m, $d = \sqrt[3]{2}\,e^2$ (m) and $D = e^{12}$ (m). What will the capacitance ratio be of the newly formed state to the previous state if the strings of 1, 2, and 3 are disconnected simultaneously from the bundle conductors so that the current passing through the strings $1'$, $2'$, $3'$ will be double that of the previous state? Do not consider the earth's effect.

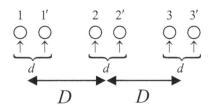

FIGURE 2.18
A horizontal three-phase (Question network 2.4).

1. $\dfrac{10}{12}$ 2. $\dfrac{11}{13}$ 3. $\dfrac{25}{26}$ 4. $\dfrac{11}{12}$

2.5. A 50-Hz single-phase transmission line has conductors with radius r and is located at a distance of re^1 (m) from each other. If the line voltage is 20 kV, what is the reactive power kVAr created by the capacitance of this line per kilometer? $(\varepsilon_0 = 9 \times 10^{-11}/\pi^2)$

    1. 0.9   2. 1.8   3. 3.6   4. 7.2

2.6. There is a telephone line d-f below the 50-Hz three-phase transmission line a-b-c in Figure 2.19. If the current in the three-phase line is 100-A per phase, how many voltages per kilometer are induced in the telephone line? Whether the line has been transposed or not, respectively.

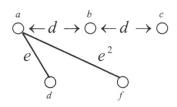

FIGURE 2.19
A horizontal three-phase (Question network 2.6).

    1. 3π and 3π 2. 2π and 0

    3. 0 and $\pi\sqrt{3}$ 4. 0 and $2\pi\sqrt{3}$

2.7. The conductors of a three-phase transmission line are at the vertices of an equilateral triangle with sides of re^1 (m) and a radius of r (m). What is the capacitance to inductance ratio of the line?

    1. $\dfrac{16\pi^2\varepsilon_0}{5\mu_0}$ 2. $\dfrac{22\pi^2\varepsilon_0}{74\mu_0}$ 3. $\dfrac{4\pi^2\varepsilon_0}{205\mu_0}$ 4. $\dfrac{8\varepsilon_0}{\pi\mu_0}$

2.8. What is the inductance of each phase of the following three-phase double circuit in terms of (mH/km) (the GMR of each conductor is $D_S = D/(2\sqrt{3})$)? (The Figure 2.20 is a regular hexagon.)

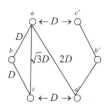

FIGURE 2.20
A regular hexagon, three-phase double circuit.

    1. $0.2\ln 3$ 2. $0.1\ln 3$ 3. $0.2\ln 4$ 4. $0.1\ln\dfrac{4}{3}$

2.9. The three-phase untransposed transmission line *a-b-c* is adjacent to the telephone line. How many times $f\mu_0$ the phase current stock of (*a*) in the induced voltage in the telephone line (Figure 2.21)?

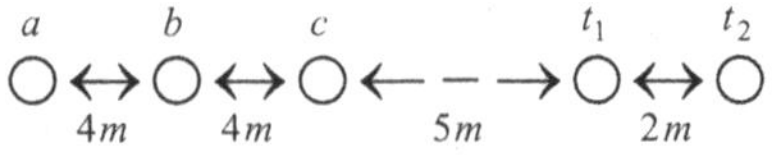

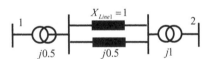

FIGURE 2.21
A three-phase untransposed transmission and a telephone line (Question network 2.9).

1. $\ln\dfrac{15}{13}$ 　　　 2. $\ln\dfrac{11}{9}$ 　　　 3. $\ln\dfrac{7}{5}$ 　　　 4. $\ln\dfrac{13}{15}$

2.10. Consider a horizontal single-phase transmission line (1-2-3). Phases (1) and (3) are sent conductors and phase (2) is the return conductor. What is the capacitance value of this single-phase line (F/m)? Do not consider the earth's effect.

1. $\dfrac{4\pi\varepsilon_0}{\ln\dfrac{d^3}{r^2}}$ 　　 2. $\dfrac{4\pi\varepsilon_0}{\ln\dfrac{2d^2}{r}}$ 　　 3. $\dfrac{4\pi\varepsilon_0}{\ln\dfrac{d^3}{2r^3}}$ 　　 4. $\dfrac{2\pi\varepsilon_0}{\ln\dfrac{d}{2r}}$

2.11. The following single-line diagram (Figure 2.22) assumes that bus voltages (1) and (2) are kept constant. The transmission line (1) is arranged into equilateral triangles with a distance of $D = re^{\frac{7}{4}}$ (m), and *r* is the radius of each conductor. How many times would the maximum transmission power increase if the distance between the phases was increased by (e^{-1}) times?

FIGURE 2.22
A single-line diagram (Question network 2.11).

1. $\dfrac{77}{24}$ 　　　 2. $\dfrac{22}{21}$ 　　　 3. 2 　　　 4. $\dfrac{22}{13}$

2.12. In a three-phase untransposed transmission line, $D_{ab}=c$, $D_{bc}=a$, $D_{ca}=b$, what is the inductance value for phase (a) of the line (H/m)?

1. $\left[1+\ln\dfrac{bc}{r}+j\sqrt{3}\ln\dfrac{b}{c}\right]\times 10^{-7}$ (H)

2. $\left[\dfrac{1}{2}+2\ln\dfrac{\sqrt{bc}}{r}+j\sqrt{3}\ln\dfrac{c}{b}\right]\times 10^{-7}$ (H)

3. $10^{-7}\ln\dfrac{(abc)^{\frac{1}{3}}}{re^{-\frac{1}{4}}}$ (H)

4. $2\times 10^{-7}\ln\dfrac{(abc)^{\frac{1}{3}}}{re^{-\frac{1}{4}}}$ (H)

2.13. Phase (a) is located at a distance (D) from another phase (b). The distance between conductors (a and b) and ground is $(H_1=H_2=H)$. Find the capacitance ratio after halving (H). The earth's effect should not be ignored.

1. $\dfrac{2\ln\left(\dfrac{D}{r\left[1+\dfrac{D^2}{H^2}\right]^{\frac{1}{2}}}\right)}{\ln\left(\dfrac{D}{r\left[1+\dfrac{D^2}{4H^2}\right]^{\frac{1}{2}}}\right)}$

2. $\dfrac{\ln\left(\dfrac{D}{1+\dfrac{D^2}{H^2}}\right)}{\ln\left(\dfrac{D}{1+\dfrac{D^2}{4H^2}}\right)}$

3. $\dfrac{2\ln\left(\dfrac{D^2}{r^2\left[1+\dfrac{D^2}{H^2}\right]^{\frac{1}{2}}}\right)}{\ln\left(\dfrac{D}{r\left[1+\dfrac{D^2}{4H^2}\right]^{\frac{1}{2}}}\right)}$

4. $\dfrac{\ln\left(\dfrac{D}{r\left[1+\dfrac{D^2}{H^2}\right]^{\frac{1}{2}}}\right)}{\ln\left(\dfrac{D}{r\left[1+\left(\dfrac{D}{2H}\right)^2\right]^{\frac{1}{2}}}\right)}$

2.14. There are two send conductors (x and y) and one return conductor (z) in a single-phase line. If the radius of each conductor is (r) and the distance between them is D, what will be the capacitance?

1. $\dfrac{2\pi\varepsilon_0}{\ln\dfrac{4D^2}{r^2}}$ 2. $\dfrac{2\pi\varepsilon_0}{\ln\dfrac{4D^3}{r}}$ 3. $\dfrac{4\pi\varepsilon_0}{\ln\dfrac{4D^2}{r^2}}$ 4. $\dfrac{4\pi\varepsilon_0}{\ln\dfrac{4D^3}{r^3}}$

2.15. Considering a three-phase double-circuit line in a regular hexagon ($A_1B_1C_1 - C_2B_2A_2$), What is the inductance of each phase in terms of (mH/km) (the GMR of each conductor is $D_S = D/(2\sqrt{3})$? (This is a regular hexagon.)

1. $0.1\ln 3$ 2. $0.1\ln 6$ 3. $0.2\ln 4$ 4. $0.1\ln 4$

2.16. Consider two, three, and four-conductor bundle. In the inductance formula, how does GMR affect each one?

1. In two-conductor bundle: $\text{GMR} \propto d^{\frac{3}{4}}$

2. In three-conductor bundle: $\text{GMR} \propto d^{\frac{2}{3}}$

3. In four-conductor bundle: $\text{GMR} \propto d^{\frac{3}{5}}$

4. In three-conductor bundle: $\text{GMR} \propto d^{-\frac{3}{4}}$

2.17. Based on the radius of each conductor (r), what is the GMR of the four conductors stuck together in a rhombus shape? (Suppose each conductor is hollow).

1. $2^{\frac{3}{4}} \cdot 3^{\frac{1}{16}} \cdot r$ 2. $2^{\frac{3}{4}} \cdot 3^{\frac{1}{8}} \cdot r$ 3. $2^{\frac{3}{2}} \cdot 3^{\frac{1}{4}} \cdot r$ 4. $2^{\frac{3}{2}} \cdot 3^{\frac{1}{16}} \cdot r$

2.18. In terms of OHM, the reactance of the horizontal three-phase with D distance is equal to π. When the distance between conductors is $D = e^3$, what is the size and type of the conductor radius? $f = 50\,\text{Hz}$, $l = 50\,\text{km}$

1. $2^{\frac{1}{3}}e^2$ and hollow   2. $2^{\frac{1}{3}}e^2$ and solid

3. $2^{\frac{1}{3}}e^{\frac{9}{4}}$ and solid   4. 1 and 3

2.19. A telephone line under a single-phase line is located in a square with a side (x). What is the induced voltage per 20-km of the telephone line? $\ln 2 = 0.7$, $\pi = 3$, $f = 50\,\text{Hz}$, $I = 50$ A, x is unknown.

1. $20\,\text{V}$ 2. $30\,\text{V}$ 3. $42\,\text{V}$ 4. $40\,\text{V}$

2.20. The send conductor (X-G) has three vertical wires that are spaced $0.5\,\text{m}$ apart, and the return conductor (Y-G) has one wire that is spaced $5\,\text{m}$ apart from the middle conductor of the send conductor.

What is the inductance (Y-G) per $10^7\,\text{m}$ in the system with hollow conductors? $r_y = 5e^{-5}\,\text{m}$

1. 10 2. 2 3. 1 4. 20

2.21. In three-phase overhead transmission lines, which arrangement has the lowest inductance (*L*) and the highest capacity (*C*) per unit length?

1. The vertical three-phase with *D* distance
2. The horizontal three-phase with *D* distance
3. Isosceles triangle, leg (*d*) is smaller than base (*D*)
4. Equilateral triangle with *D* distance

2.22. In a horizon single-phase line, the send path consists of two stranded conductors of *x* and *y*, and the return path consists of a stranded conductor of *z*. The conductors are the same and have the same GMR D_S. The distance between conductors (*x-z*) and (*z-y*) equals (*D*). How many Henrys per meter is the inductance of the single-phase line? Due to the symmetry of conductors *x* and *y* with respect to conductor *z*, the current in conductors *x* and *y* is equal.

1. $2 \times 10^{-7} \ln \dfrac{D}{D_S}$

2. $2 \times 10^{-7} \ln \sqrt{\dfrac{D_S}{2D}}$

3. $2 \times 10^{-7} \ln \sqrt{\dfrac{D}{2D_S}}$

4. $10^{-7} \left[3\ln \dfrac{D}{D_S} - \ln 2 \right]$

2.23. The GMR of four bundled conductors, which are at the corners of a square to its side (*d*), each of which has a GMR equivalent to *D*, is equal to:

1. $2^{\frac{1}{8}} D^{\frac{3}{4}} d^{\frac{3}{4}}$ 2. $4^{\frac{3}{4}} D^{\frac{3}{4}} d^{\frac{1}{4}}$ 3. $4^{\frac{1}{4}} D^{\frac{1}{4}} d^{\frac{1}{4}}$ 4. $2^{\frac{1}{8}} D^{\frac{1}{4}} d^{\frac{3}{4}}$

2.24. If we bundle a line, which of the following statements is true?

1. Bundling reduces line inductance and line capacitance but does not change the characteristic impedance of the line.
2. Bundling increases the line inductance, decreases the line capacitance, and increases the characteristic impedance of the line.
3. Bundling decreases the line inductance, increases the line capacitance, and decreases the characteristic impedance of the line.
4. Bundling increases the line inductance and the line capacitance but does not change the characteristic impedance of the line.

2.25. Three conductors (*a-b-c*) are located on an equilateral triangle with distance *D*. The single-phase transmission line has conductors (*b*) and (*c*) to send and (*a*) to return. All conductors are the same and have the same radius, *r*. The sending current from each of the sending conductors is half that of the returning current. What is the capacitance of this single-phase line?

1. $\dfrac{2\pi\varepsilon_0}{\ln\left(\dfrac{D}{r}\right)^2}$ 2. $\dfrac{2\pi\varepsilon_0}{\ln\left(\dfrac{2D}{r}\right)^3}$ 3. $\dfrac{4\pi\varepsilon_0}{\ln\left(\dfrac{D}{2r}\right)^3}$ 4. $\dfrac{4\pi\varepsilon_0}{\ln\left(\dfrac{D}{r}\right)^3}$

2.26. The resistance of an aluminum transmission line at a temperature of 20° is 10 Ω. If the loss of the line at this temperature is 11 kW at a constant load, how many kilowatts will the loss of the line be at a temperature of 40°? (Aluminum heat constant $T=200$).

1. 11   2. 15   3. 12   4. 13

2.27. On a regular hexagon with a side length of D, what is the GMD of a six-phase system?

1. $D\sqrt[6]{5}$ 2. $D^{\frac{5}{6}}3^{\frac{2}{5}}2^{\frac{1}{5}}$ 3. $D\sqrt[5]{6}$ 4. $D^{\frac{5}{6}}3^{\frac{1}{5}}2^{\frac{1}{6}}$

2.28. In a two-wire single-phase system with a length of 1 km, the power and voltage G_1 are 200 MVA and 10 kV respectively. The diameter of the hollow line is $2e^{-5}$m. The load power on bus 2 is also 1 p.u. Assuming the voltage of bus 1 and 2 remains constant at 1 p.u., how far should the two-wire be so that the generator does not lose its static stability?

1. $e^{\frac{1+0.2\pi}{0.2\pi}}$ 2. $\ln(2)\times e^{1-0.2\pi}$

3. $e^{\frac{25-5\pi}{\pi}}$ 4. $\ln(0.2)\times e^{\frac{1+0.6\pi}{0.8\pi}}$

2.29. A single-phase transmission line (a-b) and a telephone line (c-d) are installed on a tower in symmetry. What is the mutual inductance in unit length between circuits a-b and c-d?

1. $2\times 10^{-7}\ln\dfrac{D_{ad}D_{bd}}{D_{ad}D_{bc}}$ 2. $2\times 10^{-7}\ln\dfrac{D_{ad}}{D_{ac}}$

3. $4\times 10^{-7}\ln\dfrac{\sqrt{D_{ad}D_{bc}}}{\sqrt{D_{ab}D_{cd}}}$ 4. $4\times 10^{-7}\ln\dfrac{D_{ad}}{D_{ac}}$

2.30. Consider a three-phase double-circuit transmission line. The first three phases ($A_1B_1C_1$) should be arranged vertically (1 m) and the second three phases ($A_2B_2C_2$) horizontally (1 m) as in the first circuit. What is the inductance of each phase of this system in terms of (mH/km) (the GMR of each conductor is D_S)?

1. $0.2\ln\dfrac{40^{\frac{2}{3}}}{D_s^{2}}$ 2. $0.2\ln\dfrac{80^{\frac{1}{12}}}{D_S^{\frac{1}{2}}}$

3. $0.2\ln\dfrac{80^{\frac{3}{4}}}{D_s^{\frac{1}{2}}\sqrt[8]{e}}$ 4. $0.2\ln\dfrac{40^{\frac{3}{4}}}{\sqrt{D_s\cdot e}}$

2.31. A conductor (a) is located at a distance (D) from another conductor (b). The distance between conductor (a) and ground is (H_1), and the distance between conductor (b) and ground is (H_2). What is the capacitance C_{ab} with earthing effect?

1. $\dfrac{2\pi\varepsilon_0}{\ln\dfrac{2DH_1}{r\cdot\sqrt{D^2+4H_1\cdot H_2}}}$ 2. $\dfrac{\pi\varepsilon_0}{\ln\dfrac{2DH_1}{r\cdot\sqrt{D^2+4H_1\cdot H_2}}}$

3. $\dfrac{\pi\varepsilon_0}{\ln\dfrac{2D\sqrt{H_1\cdot H_2}}{r\cdot\sqrt{D^2+4H_1\cdot H_2}}}$ 4. $\dfrac{2\pi\varepsilon_0}{\ln\dfrac{2D\sqrt{H_1\cdot H_2}}{r\cdot\sqrt{D^2+4H_1\cdot H_2}}}$

2.32. What is the capacitance of one phase with radius r and with earthing effect in three horizontal a-b-c phases with distances x meters, and with earth distance x meters?

1. $\dfrac{2\pi\varepsilon_0}{\ln\left(\dfrac{x}{r}\cdot 2^{\frac{1}{3}}\right)}$ 2. $\dfrac{2\pi\varepsilon_0}{\ln\left(\dfrac{x}{r}\cdot 5^{-\frac{1}{3}}\cdot 2^{\frac{5}{6}}\right)}$

3. $\dfrac{2\pi\varepsilon_0}{\ln\left(\dfrac{x}{r}\cdot 5^{\frac{1}{3}}\cdot 2^{\frac{7}{6}}\right)}$ 4. $\dfrac{2\pi\varepsilon_0}{\ln\left(\dfrac{x}{r}\right)}$

2.33. What is the GMR of a three-circuit three-phase system? All three three-phase circuits are horizontal. The horizontal and vertical distance between the phases is (1 m).

1. $D_S^{\frac{1}{3}}\cdot 2^{\frac{1}{3}}$ 2. $D_S^{\frac{1}{9}}\cdot 2^{\frac{2}{27}}$ 3. $D_S^{\frac{2}{3}}\cdot 2^{\frac{4}{9}}$ 4. $D_S^{\frac{1}{3}}\cdot 2^{\frac{2}{9}}$

2.34. What is the GMD in question (2.33)?

1. $2^{\frac{10}{27}}\cdot 5^{\frac{4}{27}}$ 2. $2^{\frac{4}{9}}\cdot 5^{\frac{2}{9}}$ 3. $2^{\frac{5}{12}}\cdot 5^{\frac{1}{6}}$ 4. $2^{\frac{5}{9}}\cdot 5^{\frac{2}{9}}$

2.35. What is the GMD of a horizontal six-phase system (a-b-c-d-e-f) with a distance of 1 m?

1. $2^{\frac{8}{15}}\cdot 3^{\frac{1}{5}}\cdot 5^{\frac{1}{15}}$ 2. $\sqrt[25]{(2\times 3\times 4\times 5)^4}$

3. $\sqrt[5]{(2\times 3\times 4\times 5)}$ 4. $2^{\frac{5}{36}}\cdot 3^{\frac{5}{36}}\cdot 4^{\frac{5}{36}}\cdot 5^{\frac{5}{36}}$

2.8 Key Answers to Four-Choice Questions

Question	1	2	3	4
1. (2)		×		
2. (2)		×		
3. (2)		×		
4. (4)				×
5. (3)			×	
6. (4)				×
7. (1)	×			
8. (2)		×		
9. (1)	×			
10. (3)			×	
11. (2)		×		
12. (2)		×		
13. (1)	×			
14. (4)				×
15. (2)		×		
16. (2)		×		
17. (1)	×			
18. (4)				×
19. (3)			×	
20. (1)	×			
21. (3)			×	
22. (4)				×
23. (4)				×
24. (3)			×	
25. (4)				×
26. (3)			×	
27. (3)			×	
28. (3)			×	
29. (4)				×
30. (2)		×		
31. (3)			×	
32. (2)		×		
33. (4)				×
34. (1)	×			
35. (1)	×			

2.9 Descriptive Answers to Four-Choice Questions

2.1. **Option 2 is correct.** From equation (2.48) we have:

$$x = v = 1, y = \sqrt{2}, z = \sqrt{5} \Rightarrow L = 0.2 \ln \left(2^{\frac{1}{6}} \left(\frac{1}{D_S} \right)^{\frac{1}{2}} \left(\frac{2}{5} \right)^{\frac{1}{6}} \right)$$

$$= 0.2 \ln \left(2^{\frac{1}{3}} D_S^{-\frac{1}{2}} 5^{-\frac{1}{6}} \right) \text{mH/km}$$

2.2. **Option 2 is correct.** In the three-phase case:

$$L_{3ph} = 2 \times 10^{-7} \ln \frac{\text{GMD}}{\text{GMR}} \qquad D_S = \frac{d}{2}, \qquad \text{GMR} = D_S \quad \text{a string}$$

$$\text{GMD} = \sqrt[3]{d \times 2d \times d} = d\sqrt[3]{2} \Rightarrow L_{3ph} = 2 \times 10^{-7} \ln \left(2^{\frac{4}{3}} \right)$$

In the single-phase case: $\text{GMD} = \sqrt{d \times d} = d$

Send: $\text{GMR}_{x,z} = \sqrt[2^2]{D_S \times 2d \times D_S \times 2d} = \sqrt{2dD_S} = d$

Return: $\text{GMR}_y = D_S = \frac{d}{2}$

$$L_{1ph} = L_{x,z} + L_y = 2 \times 10^{-7} \left[\ln \frac{\text{GMD}}{\text{GMR}_{x,z}} + \ln \frac{\text{GMD}}{\text{GMR}_y} \right]$$

$$= 2 \times 10^{-7} \left[\ln 1 + \ln 2 \right] = 2 \times 10^{-7} \ln 2$$

$$\Rightarrow \frac{L_{3ph}}{L_{1ph}} = \frac{\ln 2^{\frac{4}{3}}}{\ln 2} = \frac{4}{3}$$

2.3. **Option 2 is correct.** Since the phase distance has not changed, $\text{GMD}_1 = \text{GMD}_2$ as a result:

$$C = \frac{2\pi\varepsilon_0}{\ln \dfrac{\text{GMD}}{\text{GMR}}} \Rightarrow C_1 = C_2 \Rightarrow \ln \frac{\text{GMD}_1}{\text{GMR}_1} = \ln \frac{\text{GMD}_2}{\text{GMR}_2} \Rightarrow \frac{\text{GMD}_1}{\text{GMR}_1} = \frac{\text{GMD}_2}{\text{GMR}_2}$$

$$\Rightarrow \text{GMR}_1 = \text{GMR}_2$$

GMR of three-conductor bundle is: $GMR_1 = \sqrt[3]{rd^2}$
GMR of two-conductor bundle is: $GMR_2 = \sqrt{rd'}$
Then:

$$\sqrt[3]{rd^2} = \sqrt{rd'} \Rightarrow r^{\frac{1}{3}} \times d^{\frac{2}{3}} = r^{\frac{1}{2}} \times d'^{\frac{1}{2}} \Rightarrow d' = d\sqrt[3]{\frac{d}{r}}$$

2.4. **Option 4 is correct.**
In the two-conductor bundle case:

$$GMD = \sqrt[3]{D \times 2D \times D} = D\sqrt[3]{2}$$

$$GMR = \sqrt[2^2]{(r \times d)^2} = \sqrt{rd} \Rightarrow C_1 = \frac{2\pi\varepsilon_0}{\ln\dfrac{D\sqrt[3]{2}}{\sqrt{rd}}}$$

In the single-string case:

$$GMD = \sqrt[3]{D \times 2D \times D} = D\sqrt[3]{2}$$

$$GMR = r = \sqrt[3]{2} \Rightarrow C_2 = \frac{2\pi\varepsilon_0}{\ln\dfrac{D\sqrt[3]{2}}{r}}$$

$$\frac{C_2}{C_1} = \frac{\ln\dfrac{\sqrt[3]{2} \times e^{12}}{\left(\sqrt[3]{2} \times \sqrt[3]{2} \times e^2\right)^{\frac{1}{2}}}}{\ln\dfrac{\sqrt[3]{2} \times e^{12}}{\sqrt[3]{2}}} = \frac{\ln e^{11}}{\ln e^{12}} = \frac{11}{12}$$

2.5. **Option 3 is correct.**
Capacitance between the send and return conductors:

$$C = \frac{\pi\varepsilon_0}{\ln\dfrac{GMD}{GMR}} = \frac{\pi\varepsilon_0}{\ln\dfrac{re^1}{r}} = \frac{\pi\varepsilon_0}{1} = \pi\varepsilon_0$$

$$Q_C = 2\pi fc V^2 \times 1\,kV = 2\pi \times 50 \times \pi \times \frac{9 \times 10^{-11}}{\pi^2} \times (20\,kV)^2 \times 1\,kV = 3.6\ kVAr/km$$

2.6. **Option 4 is correct.**
The induced voltage becomes zero if the line is transposed, and if it is untransposed:
The total flux linkage of *d-f*:

$$\lambda = 2 \times 10^{-7} \left[\hat{I}_a \ln \frac{e^2}{e} + \hat{I}_b \ln 1 + \hat{I}_c \ln \frac{e}{e^2} \right]$$

$$= 2 \times 10^{-7} \left[\hat{I}_a - \hat{I}_c \right] = 2 \times 10^{-7} \left[100\angle 0 - 100\angle 120 \right]$$

$$= 2 \times 10^{-7} \times \sqrt{3} \times 100\angle -30 \ \text{wb/m}$$

The voltage induced in the 1 km telephone line is:

$$V = \omega |\lambda| = 100\pi \times 2 \times 10^{-7} \times \sqrt{3} \times 100 \times 1000$$

$$= 2\pi \sqrt{3} \ (\text{V/km})$$

2.7. **Option 1 is correct.** Calculation of *C*:

$$GMD = re^1$$

$$GMR = r \Rightarrow C = \frac{2\pi\varepsilon_0}{\ln \dfrac{re^1}{r}} = \frac{2\pi\varepsilon_0}{1} = 2\pi\varepsilon_0 \ (\text{F/m})$$

Calculation of *L*:
Since there is no mention of solid or hollow conductors, we assume conductors to be solid. We have: $\left(2 \times 10^{-7} = \dfrac{\mu_0}{2\pi} \right)$

$$GMR = r' = 0.7788r = re^{\frac{-1}{4}}$$

$$GMD = re^1 \Rightarrow L = \frac{\mu_0}{2\pi} \ln \frac{re^1}{re^{\frac{-1}{4}}} = \frac{\mu_0}{2\pi} \times \frac{5}{4} \ (\text{H/m})$$

$$\frac{B_C}{X_L} = \frac{C\omega}{L\omega} = \frac{C}{L} = \frac{2\pi\varepsilon_0}{\dfrac{\mu_0}{2\pi} \times \dfrac{5}{4}} = \frac{16\pi^2 \varepsilon_0}{5\mu_0}$$

2.8. **Option 2 is correct.** From equation (2.46) we have:

$$L = 0.1\ln\left(\dfrac{\sqrt{3}D}{2\dfrac{D}{2\sqrt{3}}}\right) = 0.1\ln 3\,\text{mH/km}$$

2.9. **Option 1 is correct.** Based on equation (2.7) and $V = \omega\lambda$, the voltage induced by two wire ends is:

$$V = \omega \times \dfrac{\mu_0}{2\pi}\left[\hat{I}_a \ln\dfrac{15}{13} + \hat{I}_b \ln\dfrac{11}{9} + \hat{I}_c \ln\dfrac{7}{5}\right]$$

$$\text{Part } I_a = \left(f\mu_0\right) \times \ln\dfrac{15}{13}$$

2.10. **Option 3 is correct.**

The first solution:

We can calculate the send capacitance of (C_x) and the return capacitance of (C_y), and then series them together $\left(\dfrac{C_x \cdot C_y}{C_x + C_y}\right)$ or use the equation (2.36) of the potential difference between the two conductors and calculate the capacitance from the voltage, which is more appropriate due to this symmetry.

$$V_{12} = V_{23} \Rightarrow \text{Between send and return } V = \dfrac{V_{12} + V_{23}}{2} = V_{12}$$

$$V_{12} = \dfrac{1}{2\pi\varepsilon_0}\left[\dfrac{q}{2}\ln\dfrac{d}{r} - q\ln\dfrac{r}{d} + \dfrac{q}{2}\ln\dfrac{d}{2d}\right]$$

$$= \dfrac{q}{4\pi\varepsilon_0}\ln\dfrac{d^3}{2r^3} = \dfrac{q}{C_{12}} \Rightarrow C_{12} = \dfrac{4\pi\varepsilon_0}{\ln\dfrac{d^3}{2r^3}}$$

As a second solution: It is suggested to choose the answer that is closest to the general shape of capacitance means $\left(\dfrac{?}{\ln\dfrac{D}{r}}\right)$, if you don't have the time to solve the problem accurately.

2.11. **Option 2 is correct.** *The first solution:*
First case:

$$X_{12} = 0.5 + (1 \| 0.5) + 1 = \frac{11}{6}$$

$$P_{\text{max } 1} = \frac{V_1 \cdot V_2}{X_{12}} = \frac{6}{11} V_1 \cdot V_2$$

$$L_1 = 2 \times 10^{-7} \ln \frac{\text{GMD}}{\text{GMR}} = 2 \times 10^{-7} \ln \frac{re^{\frac{7}{4}}}{re^{\frac{-1}{4}}} = 4 \times 10^{-7} \ (\text{H}/\text{m})$$

Second case:

$$\left[D_2 = \frac{1}{e} D_1 = \frac{1}{e} \times re^{\frac{7}{4}} = re^{\frac{3}{4}} \right]$$

$$\Rightarrow L_2 = 2 \times 10^{-7} \ln \frac{re^{\frac{3}{4}}}{re^{\frac{-1}{4}}} = 2 \times 10^{-7} \ (\text{H}/\text{m})$$

$$X \propto L \Rightarrow L_2 = \frac{1}{2} L_1 \Rightarrow X_2 = \frac{1}{2} X_1 = 0.5$$

$$X_{12} = 0.5 + (0.5 \| 0.5) + 1 = \frac{7}{4}$$

$$P_{\text{max } 2} = \frac{V_1 \cdot V_2}{X_{12}} = \frac{4}{7} V_1 \cdot V_2$$

$$\frac{P_{\text{max } 2}}{P_{\text{max } 1}} = \frac{4/7}{6/11} = \frac{22}{21}$$

The second solution:
Since the inductance decreases with decreasing distance between the conductors, so the reactance depreciates and the transmission power increases, but compared to the reactance of series-transformers with the line, this power increase can't be large, so option 2 which is more than one and is very close to one is the correct answer.

2.12. **Option 2 is correct.** *The first solution:*

Since C and L are obtained for transposed and symmetric lines, they cannot be used here. Assuming the phase current is a positive sequence, we have:

$$\lambda_a = 2 \times 10^{-7} \left[\hat{I}_a \ln \frac{1}{r'} + \hat{I}_b \ln \frac{1}{D_{ab}} + \hat{I}_c \ln \frac{1}{D_{ca}} \right]$$

$$= 2 \times 10^{-7} \left[\hat{I}_a \ln \frac{1}{re^{\frac{-1}{4}}} + \hat{I}_a \times 1\angle 240 \times \ln \frac{1}{c} + \hat{I}_a \times 1\angle 120 \times \ln \frac{1}{b} \right]$$

$$= 2 \times 10^{-7} \hat{I}_a \left[\ln \frac{1}{re^{\frac{-1}{4}}} + \left(-0.5 - j\frac{\sqrt{3}}{2} \right) \ln \frac{1}{c} + \left(-0.5 + j\frac{\sqrt{3}}{2} \right) \ln \frac{1}{b} \right]$$

$$= 2 \times 10^{-7} \hat{I}_a \left[\ln \frac{1}{r} + \ln e^{\frac{1}{4}} - \frac{1}{2} \left[\ln \frac{1}{c} + \ln \frac{1}{b} \right] + j\frac{\sqrt{3}}{2} \left(-\ln \frac{1}{c} + \ln \frac{1}{b} \right) \right]$$

$$= 10^{-7} \hat{I}_a \left[2\ln \frac{\sqrt{bc}}{r} + \frac{1}{2} + j\sqrt{3} \ln \frac{c}{b} \right] (\text{wb/m})$$

$$\lambda_a = L_a I_a \rightarrow L_a = 10^{-7} \left[2\ln \frac{\sqrt{bc}}{r} + \frac{1}{2} + j\sqrt{3} \ln \frac{c}{b} \right] (\text{H/m})$$

The second solution: When a question has a lot of input information, it shouldn't be solved directly, but with easier conditions from the answers, the answer can be identified. In this question, since the three-phase system is an asymmetric system, answers 3 and 4 which use symmetric GMD are incorrect. For answers 1 and 2, assuming the symmetry $(a=b=c=D)$, answer 1 changes to $1 + \ln \frac{D^2}{r}$ which is different from the general relationship of inductance, so $\left(L \propto \ln \frac{D}{r} \right)$, so option 2 is correct.

2.13. **Option 1 is correct.** *The first solution:*

In the first case, (equation 2.44) can be used by assuming $H = H_1 = H_2$.

$$C_{an} = \frac{2\pi\varepsilon_0}{\ln \dfrac{D}{r\sqrt{1 + \left(\dfrac{D}{2H} \right)^2}}}$$

The second case involves the capacitance between two conductors, which is half the capacitance with respect to earth. In addition, H is changed to $H/2$.

$$C_{ab} = \frac{\pi\varepsilon_0}{\ln\dfrac{D}{r\sqrt{1+\left(\dfrac{D}{H}\right)^2}}} \Rightarrow \frac{C_{an}}{C_{ab}} = 2\frac{\ln\dfrac{D}{r\sqrt{1+\left(\dfrac{D}{H}\right)^2}}}{\ln\dfrac{D}{r\sqrt{1+\dfrac{D^2}{4H^2}}}}$$

The second solution: If $H \gg D$, the earthing effect disappears, then the ratio of *capacitance to earth* to *capacitance between conductors* must be 2. Since option 2 does not depend on r, it is incorrect. If $H \to \infty$ only option 1 tends to 2.

2.14. **Option 4 is correct.**

First, we calculate the send and return capacitance relative to the earth, and then we series them together.

$$C_{an} = \frac{2\pi\varepsilon_0}{\ln\dfrac{GMD}{GMR}} = \frac{2\pi\varepsilon_0}{\ln\dfrac{D\sqrt{2}}{\sqrt{r\cdot D}}} = \frac{4\pi\varepsilon_0}{\ln\dfrac{2D}{r}} \quad xy : \text{send}$$

$$C_{bn} = \frac{2\pi\varepsilon_0}{\ln\dfrac{D\sqrt{2}}{r}} = \frac{4\pi\varepsilon_0}{\ln\dfrac{2D^2}{r^2}} \quad z : \text{return}$$

$$\frac{1}{C_{ab}} = \frac{1}{C_{an}} + \frac{1}{C_{bn}} = \frac{\ln\dfrac{2D}{r}}{4\pi\varepsilon_0} + \frac{\ln\dfrac{2D^2}{r^2}}{4\pi\varepsilon_0} = \frac{\ln\dfrac{4D^3}{r^3}}{4\pi\varepsilon_0}$$

$$\Rightarrow C_{ab} = \frac{4\pi\varepsilon_0}{\ln\dfrac{4D^3}{r^3}}$$

2.15. **Option 2 is correct.** From equation (2.51) we have:

$$L = 0.1\ln\frac{D\sqrt{3}}{\dfrac{D}{2\sqrt{3}}} = 0.1\ln 6 \,(\text{mH/km})$$

2.16. **Option 2 is correct.** According to the explanations at the beginning of this chapter regarding the bundle of GMR, Option 2 is correct.

2.17. **Option 1 is correct.** Consider a rhombus, a side with a small diagonal $(2r)$ and a large diagonal $(2x)$. We have:

$$4r^2 = r^2 + x^2 \rightarrow x = \sqrt{3r^2} = r\sqrt{3}$$

$$\text{GMR} = \sqrt[4]{\left(r \times 2r \times 2r \times 2r\sqrt{3}\right)^2 \times \left(r \times 2r \times 2r \times 2r\right)^2} = 2^{\frac{3}{4}} \cdot 3^{\frac{1}{16}} \cdot r$$

2.18. **Option 4 is correct.**

$$X_L = 100\pi L \Rightarrow \pi = 100\pi L \Rightarrow L = 10 \text{ mH}$$

$$l = 50 \text{ km} \Rightarrow L = \frac{10}{50} = 0.2 \text{ mH/km}$$

$$L = 0.2\ln\frac{\text{GMD}}{\text{GMR}} \text{ (mH/km)} \Rightarrow 0.2 = 0.2\ln\frac{\sqrt[3]{e^3 \times e^3 \times 2e^3}}{\text{GMR}}$$

$$\ln\frac{2^{\frac{1}{3}}e^3}{\text{GMR}} = 1 \Rightarrow \frac{2^{\frac{1}{3}}e^3}{\text{GMR}} = e \Rightarrow \text{GMR} = 2^{\frac{1}{3}}e^2$$

The conductor can be hollow with a radius of $r = 2^{\frac{1}{3}}e^2$ or solid with a radius of $r = 2^{\frac{1}{3}}e^{\frac{9}{4}}$.

$$\text{GMR} = r' = 2^{\frac{1}{3}}e^{\frac{9}{4}} \cdot e^{-\frac{1}{4}} = 2^{\frac{1}{3}}e^2$$

So options 1 and 3 are correct.

2.19. **Option 3 is correct.** From equation (2.7) and $V = \lambda\omega$ we have:

$$\lambda_{t_1 t_2} = 2 \times 10^{-7}\left[\hat{I}\ln\frac{x\sqrt{2}}{x} - \hat{I}\ln\frac{x}{x\sqrt{2}}\right] = 2 \times 10^{-7} \times \hat{I} \times \ln 2 \text{ (wb/m)}$$

$$\left|V_{t_1 t_2}\right| = 2\pi f \cdot \left|\lambda_{t_1 t_2}\right| \cdot l = 100\pi \times 2 \times 10^{-7} \times 50 \times \ln 2 \times 20 \times 10^3 = 20\pi \ln 2 \approx 42$$

2.20. **Option 1 is correct.**

We ignore the effect of the distance between the send conductors in the GMD because it is ten times the distance between the send conductors. Since GMD=D, we have:

$$L = 2 \times 10^{-7}\ln\frac{\text{GMD}}{\text{GMR}} \text{ (H/m)}$$

$$\Rightarrow L = 2 \times 10^{-7} \times 10^7 \ln\frac{5}{5e^{-5}} = 2 \times 5 = 10\text{H}$$

2.21. Option 3 is correct.

For options 1 and 2 we have:

$$\text{GMD} = \sqrt[3]{D \times 2D \times D} = \sqrt[3]{2}D, \text{ and } \text{GMR} = r.$$

Therefore, these two options have equal L and C (without earthing effect)

For option 4: (Have Inductance less and C more than 1 and 2)

$$\text{GMD} = \sqrt[3]{D^3} = D, \text{GMR} = r$$

For option 3:

$$\text{GMD} = \sqrt[3]{d \times d \times D} = d^{\frac{2}{3}} \cdot D^{\frac{1}{3}} < D, \quad \text{GMR} = r$$

Therefore, option 3 has the lowest L and the highest C.

2.22. Option 4 is correct. Send (x, y) and return (z).

$$\text{Send: } L_{x,y} = 2 \times 10^{-7} \ln \frac{\sqrt{D^2}}{\sqrt{D_S \cdot 2D}} = 2 \times 10^{-7} \ln \frac{D^{\frac{1}{2}}}{(2D_S)^{\frac{1}{2}}} = 2 \times 10^{-7} \left[\frac{1}{2} \ln \frac{D}{2D_S} \right]$$

$$= 2 \times 10^{-7} \left[\frac{1}{2} \ln \frac{D}{D_S} - \frac{1}{2} \ln 2 \right]$$

$$\text{Return : } L_z = 2 \times 10^{-7} \ln \frac{D}{D_S}$$

$$L_{\text{tot}} = L_{x,y} + L_z = 2 \times 10^{-7} \left[\frac{3}{2} \ln \frac{D}{D_S} - \frac{1}{2} \ln 2 \right] = 10^{-7} \left[3 \ln \frac{D}{D_S} - \ln 2 \right]$$

2.23. Option 4 is correct.

The average radius of the string conductor is: $D_S = D$

$$\text{GMR} = \sqrt[4]{\left(D \times d \times d \times \sqrt{2}d \right)^4} = \sqrt[4]{Dd^3 \sqrt{2}}$$

$$= D^{\frac{1}{4}} \times d^{\frac{3}{4}} \times 2^{\frac{1}{8}}$$

2.24. Option 3 is correct.

$$L = 2 \times 10^{-7} \ln\frac{GMD}{GMR}, \quad C = \frac{2\pi\varepsilon_0}{\ln\dfrac{GMD}{GMR}}, \quad Z_c = \sqrt{\frac{Z}{Y}}, \quad Y \propto C, Z \propto L$$

As a result of bundling in line transmission, the GMR increases, the L decreases, the C increases, and ultimately the characteristic impedance decreases.

2.25. Option 4 is correct. We have a similar question (2.10).

$$V_{ba} = \frac{1}{2\pi\varepsilon_0}\left[-q\ln\frac{r}{D} + \frac{q}{2}\ln\frac{D}{r} + \frac{q}{2}\ln\frac{D}{D}\right] = \frac{3q}{4\pi\varepsilon_0}\ln\frac{D}{r}$$

$$C_{ba} = \frac{q}{V_{ba}} \Rightarrow C_{ba} = \frac{4\pi\varepsilon_0}{3\ln\dfrac{D}{r}} = \frac{4\pi\varepsilon_0}{\ln\left(\dfrac{D}{r}\right)^3}$$

2.26. Option 3 is correct.
The first solution: Based on equation (2.2), we have:

$$R_2 = R_1\frac{T+t_2}{T+t_1} \Rightarrow R_2 = 10\frac{200+40}{200+20} = \frac{120}{11}$$

$$I^2 = \frac{P_1}{10} = \frac{11}{10}\text{kW} = 1100 \Rightarrow P_2 = \frac{120}{11} \times 1100 = 12\,\text{kW}$$

The second solution: With increasing temperature, losses increase, so options 1 and 4 are incorrect, but not too much, so option 2 is also incorrect.

2.27. Option 3 is correct. Considering the equation (2.46) and symmetry of the figure, for the six-phase system, we have:

$$GMD = \sqrt[30]{\left(D_{ab}D_{ac}D_{ad}D_{ae}D_{af}\right)^6} = \sqrt[5]{D \times D\sqrt{3} \times 2D \times D\sqrt{3} \times D} = D\sqrt[5]{6}$$

2.28. Option 3 is correct.

$$P = \frac{V_1 \cdot V_2}{X}\sin\delta = 1 \Rightarrow \sin\delta = X, \quad Z_b = \frac{\left(10\,\text{kV}\right)^2}{200\,\text{MVA}} = 0.5\,\Omega, \quad GMR = e^{-5}$$

For static stability:

$$\delta < 90 \Rightarrow X\,\text{p.u.} < 1 \Rightarrow \frac{X(\Omega)}{Z_b(\Omega)} < 1 \Rightarrow X < 0.5\,\Omega$$

$$100\pi \times 2 \times 10^{-7} \times 10^{3} \ln \frac{GMD}{GMR} < 0.5 \Rightarrow 0.02\pi \ln \frac{GMD}{GMR} < 0.5$$

$$\ln GMD + 5 < \frac{25}{\pi} \Rightarrow \ln GMD < \frac{25 - 5\pi}{\pi} \Rightarrow GMD = D < e^{\frac{25-5\pi}{\pi}}$$

2.29. **Option 4 is correct.** Since the telephone line's current is much lower than the transmission line's current, λ_{dc} is limited to (a) and (b). From equation (2.7), we have:

$$\lambda_{dc} = 2 \times 10^{-7} \left(\hat{I} \ln \frac{D_{ad}}{D_{ac}} - \hat{I} \ln \frac{D_{bd}}{D_{bc}} \right) = \hat{I} \times 2 \times 10^{-7} \left(\ln \frac{D_{ad} D_{bc}}{D_{ac} D_{bd}} \right)$$

$$D_{ad} = D_{bc}, \quad D_{ac} = D_{bd}$$

$$\Rightarrow \lambda = \hat{I} \times 4 \times 10^{-7} \left(\ln \frac{D_{ad}}{D_{ac}} \right) \Rightarrow L = \frac{\lambda}{I} = 4 \times 10^{-7} \left(\ln \frac{D_{ad}}{D_{ac}} \right)$$

2.30. **Option 2 is correct.**

The first solution: Calculate directly:

$$D_{AB} = \sqrt[4]{D_{a_1b_1} D_{a_1b_2} D_{a_2b_1} D_{a_2b_2}} = \sqrt[4]{1 \times \sqrt{2} \times \sqrt{2} \times 1} = \sqrt[4]{2}$$

$$D_{AC} = \sqrt[4]{D_{a_1c_1} D_{a_1c_2} D_{a_2c_1} D_{a_2c_2}} = \sqrt[4]{2 \times \sqrt{5} \times \sqrt{5} \times 2} = \sqrt[4]{20}$$

$$D_{BC} = \sqrt[4]{D_{b_1c_1} D_{b_1c_2} D_{b_2c_1} D_{b_2c_2}} = \sqrt[4]{1 \times \sqrt{2} \times \sqrt{2} \times 1} = \sqrt[4]{2}$$

$$GMD = \sqrt[3]{D_{AB} D_{AC} D_{BC}} = \sqrt[3]{2^{\frac{1}{2}} \cdot 20^{\frac{1}{4}}}$$

$$D_{SA} = \sqrt{D_s D_{a_1a_2}} = \sqrt{D_s}, \quad D_{SB} = \sqrt{D_s D_{b_1b_2}} = \sqrt{D_s}, \quad D_{SC} = \sqrt{D_s D_{c_1c_2}} = \sqrt{D_s}$$

$$GMR = \sqrt[3]{D_{SA} D_{SB} D_{SC}} = D_s^{\frac{1}{2}} \Rightarrow L = 0.2 \ln \frac{GMD}{GMR} = 0.2 \ln \frac{\sqrt[3]{2^{\frac{1}{2}} \cdot 20^{\frac{1}{4}}}}{D_s^{\frac{1}{2}}}$$

$$= 0.2 \ln \frac{80^{\frac{1}{12}}}{D_s^{\frac{1}{2}}} \text{ (mH/km)}$$

The second solution: From equation (2.50), we have:

$$x = v = 1, y = \sqrt{2}, z = \sqrt{5} \Rightarrow L = 0.2 \ln \frac{2^{\frac{1}{6}} \times 1 \times 2^{\frac{1}{6}} \times 5^{\frac{1}{12}}}{D_s^{\frac{1}{2}} \times 1} = 0.2 \ln \frac{80^{\frac{1}{12}}}{D_s^{\frac{1}{2}}} \text{ (mH/km)}$$

2.31. **Option 3 is correct**. Based on equation (2.44) and the displacement of H_1 with H_2, C_{bn} can be calculated as follows:

$$\frac{1}{C_{ab}} = \frac{1}{C_{an}} + \frac{1}{C_{bn}} = \frac{\ln\left(\frac{2DH_1}{r\sqrt{D^2 + 4H_1H_2}}\right)}{2\pi\varepsilon_0} + \frac{\ln\left(\frac{2DH_2}{r\sqrt{D^2 + 4H_1H_2}}\right)}{2\pi\varepsilon_0}$$

$$= \frac{\ln\left(\frac{2D\sqrt{H_1H_2}}{r\sqrt{D^2 + 4H_1H_2}}\right)}{\pi\varepsilon_0}$$

2.32. **Option 2 is correct**. From equation (2.45), we have:

$$\text{GMR} = r, \quad \text{GMD} = \sqrt[3]{x \cdot x \cdot 2x} = x \cdot 2^{\frac{1}{3}}, \quad H = 2 \cdot \sqrt[3]{x \cdot x \cdot x} = 2 \cdot x$$

$$H_S = \sqrt[3]{\left(\sqrt{x^2 + 4x^2}\right)^2 \left(2x\sqrt{2}\right)} = x\sqrt[3]{5 \times 2\sqrt{2}} = x \cdot 5^{\frac{1}{3}} \cdot 2^{\frac{1}{2}}$$

$$C = \frac{2\pi\varepsilon_0}{\ln\left(\frac{x \cdot 2^{\frac{1}{3}} \cdot 2x}{r \cdot x \cdot 5^{\frac{1}{3}} \cdot 2^{\frac{1}{2}}}\right)} = \frac{2\pi\varepsilon_0}{\ln\left(\frac{x}{r} \cdot 2^{\frac{5}{6}} \cdot 5^{-\frac{1}{3}}\right)}$$

2.33. **Option 4 is correct**.

$$\text{GMR} = \sqrt[3]{\text{GMR}_a \text{GMR}_b \text{GMR}_c}$$

$$\text{GMR}_a = \text{GMR}_b = \text{GMR}_c = \sqrt[9]{D_S^3 \times 1 \times 2 \times 1 \times 1 \times 1 \times 2} = D_S^{\frac{1}{3}} \cdot 2^{\frac{2}{9}}$$

2.34. **Option 1 is correct**.

$$\text{GMD} = \sqrt[3]{D_{AB} \cdot D_{AC} \cdot D_{BC}}$$

$$D_{AB} = \sqrt[9]{D_{a1b1} \cdot D_{a1b2} \cdot D_{a1b3} \cdot D_{a2b1} \cdot D_{a2b2} \cdot D_{a2b3} \cdot D_{a3b1} \cdot D_{a3b2} \cdot D_{a3b3}}$$

$$D_{AB} = \sqrt[9]{1 \times \sqrt{2} \times \sqrt{5} \times \sqrt{2} \times 1 \times \sqrt{2} \times \sqrt{5} \times \sqrt{2} \times 1} = \sqrt[9]{2^2 \times 5} = 2^{\frac{2}{9}} \cdot 5^{\frac{1}{9}} = D_{BC}$$

$$D_{AC} = \sqrt[9]{2 \times \sqrt{5} \times 2\sqrt{2} \times \sqrt{5} \times 2 \times \sqrt{5} \times 2\sqrt{2} \times \sqrt{5} \times 2} = \sqrt[9]{2^6 \cdot 5^2} = 2^{\frac{6}{9}} \cdot 5^{\frac{2}{9}}$$

$$\Rightarrow \text{GMD} = \left(2^{\frac{4}{9}} \cdot 5^{\frac{2}{9}} \cdot 2^{\frac{6}{9}} \cdot 5^{\frac{2}{9}}\right)^{\frac{1}{3}} = 2^{\frac{10}{27}} \cdot 5^{\frac{4}{27}}$$

2.35. **Option 1 is correct.** From the main relation (2.13), we can calculate the GMD of a six-phase system:

$$GMD_{6\,ph} = \sqrt[15]{D_{ab} \cdot D_{ac} \cdot D_{ad} \cdot D_{ae} \cdot D_{af} \cdot D_{bc} \cdot D_{bd} \cdot D_{be} \cdot D_{bf} \cdot D_{cd} \cdot D_{ce} \cdot D_{cf} \cdot D_{de} \cdot D_{df} \cdot D_{ef}}$$

$$GMD_{6\,ph} = \sqrt[15]{1 \times 2 \times 3 \times 4 \times 5 \times 1 \times 2 \times 3 \times 4 \times 1 \times 2 \times 3 \times 1 \times 2 \times 1} = 2^{\frac{8}{15}} \cdot 3^{\frac{1}{5}} \cdot 5^{\frac{1}{15}}$$

2.10 Two-Choice Questions (Yes – No) – 80 Questions

1. A transmission network's main purpose is to transfer electrical energy from generators to loads.
2. A magnetic field and an electrical field affect the conductor's inductance and capacitance, respectively.
3. Leakage currents from insulators and ionized pathways in the air are shown by parallel conductivity.
4. Show the geometric mean radius and geometric mean distance in transmission lines using GMD and GMR, respectively.
5. Choosing an economic voltage level for a transmission line depends on the amount of transmission power and the distance over which it must be transmitted.
6. For flexibility, line conductors are made of stringy material.
7. Bundled conductors reduce the reactance of the line more than a new line.
8. With voltages above 230 kV, it is preferred to use more than one conductor per phase.
9. Bundling reduces the effective radius of the conductor of the line.
10. Bundling increases the force of an electrical field in the near of conductors.
11. Corona losses are reduced by bundling but radio disturbances are increased.
12. Bundling increases the reactance of the line.
13. The twisted conductors of the line have no effect on their resistance.
14. Conductors' resistance decreases due to the skin effect.
15. Current density at the surface of the conductor increases as frequency increases.
16. The actual length of each string in a twisted line is less than the length of the conductor.

17. In a single-phase system, each phase's inductance equals:

$$L = 0.2\ln\frac{D}{D_S}\ (\text{H/m})$$

18. The inductance formula of a phase in a three-phase circuit with equal distances is equal to the inductance formula of a conductor in a single-phase circuit.

19. The voltage drop on a three-phase line with unequal intervals is balanced if the currents are balanced.

20. Displacing the line conductors is an appropriate method for creating symmetry and modeling each phase.

21. The line is transposed if only phases (a) and (b) are swapped.

22. Bundling increases the transmission capacity of line power.

23. Bundling reduces the voltage gradient at the surface of the conductor.

24. Bundling increases the wave impedance of the line.

25. An electrical system's capacitance between conductors is determined by their distance from the earth.

26. In a single-phase system, a solid conductor's capacitance is equal to

$$C = \frac{2\pi\varepsilon_0}{\ln\dfrac{d}{0.7788r}}.$$

27. Bundling converts the conductor radius (r) to the equivalent radius of D_S.

28. Earth effects increase the capacitance of a line.

29. The earth effect reduces the line charging current of a transmission line.

30. The density of current induced in the human body by electrical fields of transmission lines is much lower than that induced by magnetic fields.

31. The main cause of voltage induction in telephone lines parallel to transmission lines is electric fields.

32. Corona is the ionization of air surrounding a conductor.

33. Corona refers to the failure of the dielectric in the air surrounding the conductor.

34. Coronas increase transmission line losses.

35. Corona is a function of the arrangement of lines.

36. Pollution particles on the conductor surface increase the Corona.

37. Bundling increases transmission capacity more than constructing a new line.

38. By reducing the cross-section of the conductor, the Corona is reduced.

39. It is assumed that the conductance of overhead lines is infinite because the leakage current in overhead line insulators is negligible.

40. An impedance series is formed by the resistance and inductance that form uniformly along the line.

41. An admittance parallel is formed by the conductance and capacitance that form uniformly along the line.

42. The series and parallel reactance of a line are equal when the line length is a quarter of the wavelength.

43. Power loss in a transmission line is primarily caused by the transmission line's resistance.

44. In a uniform current distribution, the effective resistance is the same as the DC resistance.

45. By increasing the frequency, the current density in the conductor becomes non-uniform.

46. Current density usually decreases from the center of a circular conductor to its surface.

47. A greater induction of voltage in the internal strings will decrease the conductor's effective resistance.

48. Inductance is the flux around a conductor required to pass a current of 1 A.

49. As a result, the capacitance between the conductors is modeled as a series capacitance.

50. Because the ground affects the electric field of the line, the capacitance of the line changes.

51. By ignoring the bundling effect, we calculate the GMD of the three-phase system to calculate capacitance.

52. The earth reduces the capacitance of transmission lines.

53. The Corona effect is shown by the parallel conductance of a line.

54. At a high voltage, a Corona occurs.

55. The primary reason for bundling is to minimize Corona effects.

56. Solid conductors have a fixed internal inductance.

57. Inductance is not affected by the conductor's shape, whether it is solid or hollow.

58. In a single-phase system, the GMD doubles if the distance between the send and return conductors doubles.

59. In a single-phase system, doubling the distance between the send and return conductors does not change the GMR.

60. In a transposed system, the mutual inductance is zero.

61. The GMR of a four-bundel conductor is in the form of $\sqrt[4]{D_S 2d^3}$.

62. Inductance does not change if the GMD is (e^2) times and the GMR is (e) times.

63. The GMR can be approximated if the distance between the send and return conductors is wider than the distance between the send and return conductors.

64. When three-phase lines are transposed and the system is unbalanced, the induced voltage in the telephone lines does not become zero.

65. For a six-phase system, 16 distances are needed to calculate the GMD.

66. A two-circuit three-phase system and a six-phase system have similar GMDs.

67. The amount of ε_0 is 8.85×10^{-12} F/m.

68. GMR in the capacitance formula uses D_S.

69. Single-phase capacitance to earth is twice that of phase-to-phase capacitance.

70. A line charging current is the current flowing through the line when there is no load on the line.

71. A three-phase system is calculated with one capacitance if the distances between the phases are not equal but the system has been transposed.

72. From the perspective of power system engineers, the inductance of the line is the most critical parameter.

73. The main reason for using aluminum in the lines instead of copper is that aluminum has a larger diameter than copper of the same weight.

74. Cables with alternating current cannot be used for long distances due to leakage current due to high reactance.

75. A short circuit in the system may result from lines warming up.

76. Inability to calculate accurately is one reason for ignoring the parallel conductance of the line.

77. The earth effect affects the magnetic field of the line.

78. Laplace's differential equation proves that earth effects are considered by considering the load's image in the earth.

79. L and C are calculated assuming the line length is limited.

80. The relation $\dfrac{1}{2\pi\varepsilon_0} = 18 \times 10^9$ is correct.

2.11 Key Answers to Two-Choice Questions

1. Yes
2. No
3. Yes
4. No
5. Yes
6. Yes
7. No
8. Yes
9. No
10. No
11. No
12. No
13. No
14. No
15. Yes
16. No
17. No
18. Yes
19. No
20. Yes
21. No
22. Yes
23. Yes
24. No
25. Yes
26. No
27. Yes
28. Yes
29. No
30. No
31. Yes
32. Yes
33. Yes

34. Yes
35. Yes
36. Yes
37. No
38. No
39. No
40. Yes
41. Yes
42. Yes
43. Yes
44. Yes
45. Yes
46. No
47. No
48. Yes
49. No
50. Yes
51. Yes
52. No
53. Yes
54. No
55. Yes
56. Yes
57. No
58. Yes
59. Yes
60. Yes
61. No
62. No
63. No
64. Yes
65. No
66. No
67. Yes
68. No
69. Yes

70. Yes
71. Yes
72. Yes
73. Yes
74. No
75. Yes
76. Yes
77. No
78. Yes
79. No
80. Yes

3

Line Model and Performance

Part One: Lesson Summary

3.1 Introduction

An accurate and appropriate model of the transmission lines is necessary to load flow of the power system. It is only the electrical model of the lines that is analyzed and reviewed in this chapter, not their mechanical design.

As a result, we have dedicated a separate chapter to the transmission line's model and performance. This chapter focuses on:

Models of transmission lines (π and T), parameters of transmission lines, classification of transmission lines (short, medium, and long), conversion of different models of lines to each other, regulation of voltage, short circuit and open-circuit impedance, ABCD parameters of transmission lines, mode special networks, voltage and current traveling waves, lossless line, wave impedance loading (SIL), line compensation, charging current in transmission line, and maximum power transmission.

This chapter emphasizes various points and requires the reader to read this lesson summary before answering the questions.

3.2 Transmission Line Models

Each line is represented by a T transmission matrix as follows:

$$\begin{bmatrix} \hat{V}_S \\ \hat{I}_S \end{bmatrix} = \underbrace{\begin{bmatrix} A & B \\ C & D \end{bmatrix}}_{T} \begin{bmatrix} \hat{V}_R \\ \hat{I}_R \end{bmatrix} \tag{3.1}$$

$$\hat{V}_S = A \cdot \hat{V}_R + B \cdot \hat{I}_R$$

$$\hat{I}_S = C \cdot \hat{V}_R + D \cdot \hat{I}_R$$

The source voltage and current are $\hat{V}_S$ and $\hat{I}_S$, and the load voltage and current are $\hat{V}_R$ and $\hat{I}_R$.

The symmetric T matrix has the following properties: $A = D, \quad A^2 - BC = 1$.

Note that the transmission model shown is correct for a single-phase system or a per-unit three-phase system. Each phase of a three-phase system must be based on the same model.

3.2.1 Displacement of Input and Output

$$
\begin{bmatrix} \hat{V}_R \\ \hat{I}_R \end{bmatrix} = \underbrace{\begin{bmatrix} D & -B \\ -C & A \end{bmatrix}}_{T^{-1}} \begin{bmatrix} \hat{V}_S \\ \hat{I}_S \end{bmatrix}
\tag{3.2}
$$

3.3 Transmission Line Parameters

These are: $z, r, x, y, g, b, l, Z, R, X, Y, G, B, L,$ and C that are defined as follows (Table 3.1):

TABLE 3.1

The ANCE Parameters Table

$Z = R + jX\ (\Omega)$ is the per-phase series **Impedance**	$z = r + jx\ (\Omega/\text{m})$ is the per-phase series impedance per unit length
$R(\Omega)$ is the per-phase series **Resistance**	$r(\Omega/\text{m})$ is the per-phase series resistance per unit length
$X(\Omega)$ is the per-phase series **Reactance**	$x\ (\Omega/\text{m})$ is the per-phase series reactance per unit length
$Y = G + jB\ (\mho)$ is the per-phase shunt **Admittance**	$y = g + jb\ (\mho/\text{m})$ is the per-phase shunt admittance per unit length
$G(\mho)$ is the per-phase shunt **Conductance**	$g\ (\mho/\text{m})$ is the per-phase shunt conductance per unit length
$B\ (\mho)$ is the per-phase shunt **Susceptance**	$b\ (\mho/\text{m})$ is the per-phase shunt susceptance per unit length
l is the line length $(m),\ (km)$ or $(mile)$ *(Make sure the rest of the parameters match)*	$L\ (\text{H}/\text{m})$ is the per-phase series inductance per unit length
	$C\ (\text{F}/\text{m})$ is the per-phase shunt capacitance per unit length

$L = \dfrac{N^2}{\mathfrak{R}}$, N is the number of turns in a coil and $\mathfrak{R}$ is the **Reluctance** and $\mathbb{P} = \dfrac{1}{\mathfrak{R}}$ is **Permeance**

Notes:

- It is imperative to remember that series impedance (Z or z) is not the inverse of shunt admittance (Y or y). $Z_{series} \neq \dfrac{1}{Y_{shunt}}$

- Generally, the (x) parameter is used to specify an unknown parameter, such as the desired distance from the beginning of the line.

3.4 Classification of Transmission Lines

a. Short-line $l < 80 \text{ km} \left(= 50 \text{ miles} \right)$

b. Medium-line $80 \text{ km} < l < 240 \text{ km} \left(= 150 \text{ miles} \right)$

c. Long-line $l > 240 \text{ km}$

Note:

- For some of the reasons outlined in the question, it might be necessary to use models with different lengths. Therefore, the model mentioned in the question would be preferable to the above division.

3.4.1 Short-Line Model

This model ignores the shunt Y:

$$\begin{bmatrix} \hat{V}_S \\ \hat{I}_S \end{bmatrix} = \underbrace{\begin{bmatrix} 1 & Z \\ 0 & 1 \end{bmatrix}}_{T} \begin{bmatrix} \hat{V}_R \\ \hat{I}_R \end{bmatrix} \tag{3.3}$$

3.4.2 Medium-Line Model

Two types of symmetric π and T are used in this model:
Symmetric π model (Figure 3.1):

$$T = \begin{bmatrix} 1 + \dfrac{ZY}{2} & Z \\ Y \left(1 + \dfrac{ZY}{4} \right) & 1 + \dfrac{ZY}{2} \end{bmatrix} \tag{3.4}$$

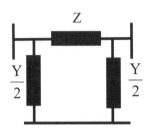

FIGURE 3.1
Symmetric π model.

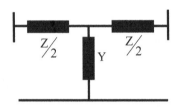

FIGURE 3.2
Symmetric *T* model.

Symmetric *T* model (Figure 3.2):

$$T = \begin{bmatrix} 1+\dfrac{ZY}{2} & Z\left(1+\dfrac{ZY}{4}\right) \\ Y & 1+\dfrac{ZY}{2} \end{bmatrix} \tag{3.5}$$

3.4.3 Long-Line Model

When the voltage and current are considered at a distance x from the load:

$$T = \begin{bmatrix} \cosh(\gamma x) & Z_C \sinh(\gamma x) \\ \dfrac{\sinh(\gamma x)}{Z_C} & \cosh(\gamma x) \end{bmatrix} \tag{3.6}$$

Transmission matrix for a long-length transmission line with $x=l$:

$$T = \begin{bmatrix} \cosh(\gamma l) & Z_C \sinh(\gamma l) \\ \dfrac{\sinh(\gamma l)}{Z_C} & \cosh(\gamma l) \end{bmatrix} \tag{3.7}$$

The propagation constant $(1/m)$:

$$\gamma \triangleq \alpha + j\beta = \sqrt{z \cdot y} = \sqrt{|z| \cdot |y|} \angle \left(\frac{\varphi_z + \varphi_y}{2} \right) \tag{3.8}$$

γl without units:

$$\gamma l = \sqrt{Z \cdot Y} = \sqrt{|Z| \cdot |Y|} \angle \left(\frac{\varphi_z + \varphi_y}{2} \right) \tag{3.9}$$

The characteristic impedance (Ω).

$$Z_C \triangleq \sqrt{\frac{z}{y}} = \sqrt{\frac{Z}{Y}} = \sqrt{\frac{|Z|}{|Y|}} \angle \left(\frac{\varphi_Z - \varphi_Y}{2} \right) \tag{3.10}$$

3.4.3.1 Hyperbolic Relationships: (x Can Be Complex)

$$\cosh x = \frac{e^x + e^{-x}}{2} \tag{3.11}$$

$$\sinh x = \frac{e^x - e^{-x}}{2}, \quad \tanh x = \frac{\sinh x}{\cosh x}, \quad (\cosh x)^2 - (\sinh x)^2 = 1$$

$$\sinh x + \cosh x = e^x$$

$$\cosh(\alpha + j\beta) = \cosh \alpha \cdot \cos(\beta) + j \sinh \alpha \cdot \sin \beta \Rightarrow \cosh(j\beta) = \cos \beta$$

$$\sinh(\alpha + j\beta) = \sinh \alpha \cdot \cos(\beta) + j \cosh \alpha \cdot \sin \beta \Rightarrow \sinh(j\beta) = j \sin \beta$$

$$\cosh(2x) = 2(\sinh x)^2 + 1, \quad \sinh(2x) = 2 \sinh x \cdot \cosh x$$

3.5 Convert Different Line Models to Each Other

3.5.1 Convert a Long-Line Model to the Similar π Model

$$Y' = Y \cdot \frac{\tanh\left(\frac{\gamma l}{2}\right)}{\frac{\gamma l}{2}} \quad \text{and} \quad Z' = Z \cdot \frac{\sinh(\gamma l)}{\gamma l} \tag{3.12}$$

3.5.2 Convert a Long-Line Model to a Similar *T* Model

$$Z' = Z \cdot \frac{\tanh\left(\dfrac{\gamma l}{2}\right)}{\dfrac{\gamma l}{2}} \quad \text{and} \quad Y' = Y \cdot \frac{\sinh(\gamma l)}{\gamma l} \tag{3.13}$$

Note:
 If $\gamma l \to 0$ then $\sinh(\gamma l) \to \gamma l$ and $\tanh(\gamma l) \to \gamma l$ so $Z' \to Z$ and $Y' \to Y$

3.5.3 Convert a Long-Line Model to Power Series

$$\cosh(\gamma l) = 1 + \frac{(\gamma l)^2}{2!} + \frac{(\gamma l)^4}{4!} + \cdots = 1 + \frac{(ZY)}{2!} + \frac{(ZY)^2}{4!} + \cdots \tag{3.14}$$

$$\sinh(\gamma l) = \gamma l + \frac{(\gamma l)^3}{3!} + \frac{(\gamma l)^5}{5!} + \cdots = \sqrt{ZY}\left(1 + \frac{(ZY)}{3!} + \frac{(ZY)^2}{5!} + \cdots\right)$$

$$A = D = \cosh(\gamma l) \approx 1 + \frac{ZY}{2}$$

$$B = Z_C \cdot \sinh(\gamma l) = Z\left(1 + \frac{ZY}{6}\right)$$

$$C = \frac{\sinh(\gamma l)}{Z_C} = Y\left(1 + \frac{ZY}{6}\right)$$

3.5.4 Two Series Transmission Lines Model (Cascade)

$$\begin{bmatrix} A & B \\ C & D \end{bmatrix} = \begin{bmatrix} A_1 & B_1 \\ C_1 & D_1 \end{bmatrix} \begin{bmatrix} A_2 & B_2 \\ C_2 & D_2 \end{bmatrix} \tag{3.15}$$

3.5.5 Two Parallel Transmission Lines Model

$$A = \frac{A_1 B_2 + B_1 A_2}{B_1 + B_2} \quad B = \frac{B_1 \cdot B_2}{B_1 + B_2} \tag{3.16}$$

$$C = C_1 + C_2 + \frac{(A_1 - A_2)(D_2 - D_1)}{B_1 + B_2} \quad D = \frac{B_1 D_2 + B_2 D_1}{B_1 + B_2}$$

The model of two parallel lines is similar to each other:

$$\begin{bmatrix} A & B \\ C & D \end{bmatrix} = \begin{bmatrix} A_1 & \dfrac{B_1}{2} \\ 2C_1 & D_1 \end{bmatrix} \tag{3.17}$$

The model of n parallel lines is similar to each other:

$$\begin{bmatrix} A & B \\ C & D \end{bmatrix} = \begin{bmatrix} A_1 & \dfrac{B_1}{n} \\ nC_1 & D_1 \end{bmatrix} \tag{3.18}$$

3.5.6 One-Half of the Transmission Line Model

Suppose $(ABCD)$ is the constant of the line to length l and $(abcd)$ is the constant of the line to length $0.5l$.

$$b = \frac{B}{2a} \text{ and } c = \frac{C}{2a} \text{ and } a = d = \sqrt{\frac{1+A}{2}} \tag{3.19}$$

3.5.7 Transmission Matrix of Series Impedance Z (Similar to Short Line Model)

$$T = \begin{bmatrix} 1 & Z \\ 0 & 1 \end{bmatrix} \tag{3.20}$$

3.5.8 Transmission Matrix of Shunt Admittance Y

$$T = \begin{bmatrix} 1 & 0 \\ Y & 1 \end{bmatrix} \tag{3.21}$$

3.6 The Voltage Regulation or Regulation ($R\%$, Reg%, or $V_R\%$)

(A) model is usually used.

A. Assume V_S is constant:

$$R\% = \frac{V_{Rnl} - V_{Rfl}}{V_{Rfl}} \times 100 \tag{3.22}$$

That:

V_{Rnl} is the voltage at the load when the load is disconnected, while V_{Rfl} is the voltage at the load when the load is connected.

B. Assume V_R is constant:

$$R\% = \frac{V_{Sfl} - V_{Snl}}{V_{Snl}} \times 100 \qquad (3.23)$$

V_{Snl} is the voltage at the source when the load is disconnected, while V_{Sfl} is the voltage at the source when the load is connected.

3.6.1 Voltage Regulation, Approximate Relation for Short Line

P_R and Q_R represent the power consumed by the load.

$$R\% = \frac{R \cdot P_R + X \cdot Q_R}{V_R^2} \times 100 \qquad (3.24)$$

3.6.2 Voltage Regulation, All Lines

$$R\% = \frac{\dfrac{V_S}{|A|} - V_{Rfl}}{V_{Rfl}} \times 100 \qquad (3.25)$$

3.6.3 Percentage of Voltage Drop

$$\Delta V\% = \frac{V_S - V_R}{V_S} \times 100 \qquad (3.26)$$

3.7 The Short-Circuit and the Open-Circuit Impedance

Suppose Z_{SS} and Z_{SO} are the short-circuit impedance and open-circuit impedance seen at the source, respectively, and can be expressed as follows:

$$Z_{SO} = \frac{\hat{V}_{SO}}{\hat{I}_{SO}} = \frac{A}{C}, \quad Z_{SS} = \frac{\hat{V}_{SS}}{\hat{I}_{SS}} = \frac{B}{D} = \frac{B}{A} \qquad (3.27)$$

$$A = D = \sqrt{\frac{Z_{SO}}{Z_{SO} - Z_{SS}}}, \quad B = Z_{SS} \cdot A, \quad C = \frac{A}{Z_{SO}}$$

In the long line: $Z_C = \sqrt{Z_{SO} \cdot Z_{SS}}$

3.8 ABCD Parameters of Transmission Lines

$$A = D = \frac{\hat{V}_s \cdot \hat{I}_s + \hat{V}_r \cdot \hat{I}_r}{\hat{V}_r \cdot \hat{I}_s + \hat{V}_s \cdot \hat{I}_r}, \quad B = \frac{\hat{V}_s^2 - \hat{V}_r^2}{\hat{V}_r \cdot \hat{I}_s + \hat{V}_s \cdot \hat{I}_r}, \quad C = \frac{\hat{I}_s^2 - \hat{I}_r^2}{\hat{V}_r \cdot \hat{I}_s + \hat{V}_s \cdot \hat{I}_r} \qquad (3.28)$$

3.9 Special Modes

A. Half the *T* model (Figure 3.3)

$$T = \begin{bmatrix} 1 + ZY & Z \\ Y & 1 \end{bmatrix} \qquad (3.29)$$

B. Asymmetric π model (Figure 3.4)

$$T = \begin{bmatrix} 1 + ZY_R & Z \\ Y_S + Y_R + Y_S Y_R Z & 1 + ZY_S \end{bmatrix} \qquad (3.30)$$

FIGURE 3.3
Unsymmetrical half-*T* model.

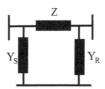

FIGURE 3.4
Unsymmetrical π model.

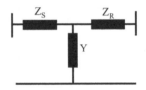

FIGURE 3.5
Unsymmetrical T model.

C. Asymmetric T model (Figure 3.5)

$$T = \begin{bmatrix} 1 + Z_S Y & Z_S + Z_R + Z_S Z_R Y \\ Y & 1 + Z_R Y \end{bmatrix} \tag{3.31}$$

3.10 Traveling Waves of Voltage and Current

Traveling voltage wave (x=distance from the beginning of the line)

$$v(t, x) = v_1(t, x) + v_2(t, x)$$

Incident wave: $v_1(t, x) = \sqrt{2} A_1 \cdot e^{\alpha x} \cdot \cos(\omega t + \beta x)$

Reflected wave: $v_2(t, x) = \sqrt{2} A_2 \cdot e^{-\alpha x} \cdot \cos(\omega t - \beta x)$ $\tag{3.32}$

Based on equation (3.8), α, A_1, β, and A_2, are determined:

$$A_1 = \left| \frac{\hat{V}_R + Z_C \hat{I}_R}{2} \right| \quad \text{and} \quad A_2 = \left| \frac{\hat{V}_R - Z_C \hat{I}_R}{2} \right| \tag{3.33}$$

Note:

- Current and voltage have the same relationship.

3.11 Lossless Line

$$r = 0, \quad g = 0 \Rightarrow \quad \alpha = 0 \Rightarrow \gamma = j\beta = j\omega\sqrt{L \cdot C} \tag{3.34}$$

$$\Rightarrow Z_C = \sqrt{\frac{L}{C}} \overset{*}{=} \frac{1}{2\pi}\sqrt{\frac{\mu_0}{\varepsilon_0}} \ln\left(\frac{\text{GMD}}{\text{GMR}}\right) = 60 \ln\left(\frac{\text{GMD}}{\text{GMR}}\right) \tag{3.35}$$

$$\Rightarrow \overset{*}{\beta} = \omega\sqrt{\mu_0 \cdot \varepsilon_0} = \frac{\omega}{3} \times 10^{-8} \tag{3.36}$$

*: Neglecting the internal flux linkage of a conductor. $GMR_L = GMR_C$.
The velocity of propagation:

$$v = \frac{\omega}{\beta} = \frac{\omega}{\omega\sqrt{L \cdot C}} = \frac{1}{\sqrt{L \cdot C}} \overset{*}{=} \frac{1}{\sqrt{\mu_0 \cdot \varepsilon_0}} = 3 \times 10^8 \text{ m/s} \tag{3.37}$$

Wavelength:

$$\lambda = \frac{2\pi}{\beta} = \frac{1}{f\sqrt{L \cdot C}} \overset{*}{=} \frac{3 \times 10^8 \text{ (m)}}{f} \tag{3.38}$$

3.11.1 Surge Load, Natural Load, Surge Impedance Loading

When $Z_{Load} = Z_C$, the loading is known as the natural load or surge impedance load (SIL).
  In a lossless line, power is delivered by its surge impedance:

$$SIL = 3 \cdot \hat{V}_R \cdot \hat{I}_R^* = \frac{3 \cdot V_R^2}{Z_C} = P + j0 \tag{3.39}$$

$$Z_C = Z_R = Z_S = \frac{\hat{V}_S}{\hat{I}_S} = \frac{\hat{V}_R}{\hat{I}_R} = \sqrt{\frac{L}{C}} = 60\ln\left(\frac{GMD}{GMR}\right) \tag{3.40}$$

There is no reactive power in the line:

$$Q_R = Q_S = 0 \tag{3.41}$$

Inductance absorbs the same reactive power as capacitance generates.

$$\omega \cdot L \cdot I_R^2 = \omega \cdot C \cdot V_R^2 \tag{3.42}$$

The magnitude of the voltage and current at each point of the line is constant.

$$\hat{V}_x = e^{j\beta x} \cdot \hat{V}_R \Rightarrow V_x = V_R, \quad \hat{I}_x = e^{j\beta x} \cdot \hat{I}_R \Rightarrow I_x = I_R \tag{3.43}$$

3.11.2 Typical Value of Z_c and *SIL*, Single Circuit

  No bundles, $Z_C = 400\,\Omega$/phase, voltage 132 kV: SIL = 43.56 MW
  Two-bundles: $Z_C = 320\,\Omega$/phase, voltage 230 kV: SIL = 165.3 MW
  Three-bundles: $Z_C = 280\,\Omega$/phase, voltage 230 kV: SIL = 189 MW
  Four-bundles: $Z_C = 260\,\Omega$/phase, voltage 400 kV: SIL = 615 MW

3.11.3 Complex Power Flow in Lossless Line

$$P_{3\,ph} = \frac{V_{SL-L} \cdot V_{RL-L}}{X'} \sin(\delta), \quad X' = Z_C \cdot \sin(\beta l) \qquad (3.44)$$

$$Q_{R3\,ph} = \frac{V_{SL-L} \cdot V_{RL-L}}{X'} \cos(\delta) - \frac{V_{RL-L}^2}{X'} \cos(\beta l) \qquad (3.45)$$

3.11.4 Power Transmission Capability and SIL in Lossless Line

$$P_{3\,ph} = \frac{V_S(\text{p.u.}) \cdot V_R(\text{p.u.}) \cdot \text{SIL}}{\sin\left(\dfrac{2\pi l}{\lambda}\right)} \sin(\delta) = \frac{V_S(\text{p.u.}) \cdot V_R(\text{p.u.}) \cdot \text{SIL}}{\sin(\beta l)} \sin(\delta) \quad (3.46)$$

3.12 Line Compensation, Shunt Reactor

Shunt reactors are used to remove the effects of line capacitance on long lines. Assuming X_{Lsh} is connected at the receiving end, we have:

$$X_{Lsh} = \frac{\sin(\beta l)}{\dfrac{V_S}{V_R} - \cos(\beta l)} Z_C \qquad (3.47)$$

The phase of V_S and V_R are equal in this case, so their magnitude can be used instead of the phasor.

The Shunt reactor for $V_S = V_R$:

$$X_{Lsh} = \frac{\sin(\beta l)}{1 - \cos(\beta l)} Z_C \qquad (3.48)$$

When we place X_{Lsh} at the end of the line, we get:

$$I_S = -I_R, \quad V_m = \frac{V_R}{\cos\left(\dfrac{\beta l}{2}\right)}, \quad I_m = 0 \qquad (3.49)$$

In other words, V_m and I_m are the voltage and current in the middle of the line, respectively.

Percentage compensation for the transmission line

Typically between 25% and 75%:

$$\frac{Y_{Lsh}}{Y_c} \tag{3.50}$$

That:
Y_{Lsh} is an admittance of compensation with a shunt reactor.
Y_c is the transmission line shunt admittance in the π model.

3.12.1 Line Compensation, Shunt Capacitor

Shunt capacitors are used in distribution networks to modify the power factor and increase voltage.

3.12.2 Line Compensation, Series Capacitor

Usually, the capacitor is placed in the middle of the line to increase power in the transmission line, but it could cause resonance during a short circuit. In this case, the power transfer in the transmission line is equal to:

$$P_{3ph} = \frac{V_{SL-L} \cdot V_{RL-L}}{X' - X_{Cser}} \sin(\delta) \tag{3.51}$$

3.12.2.1 Percentage Compensation for the Transmission Line

Typically between 25% and 75%:

$$\frac{X_C}{X'} \tag{3.52}$$

3.12.2.2 The Sub-Synchronous Resonance Frequency

$$f_r = \frac{f_s}{\sqrt{L' \cdot C_{ser}}} \tag{3.53}$$

That
f_s is the synchronous frequency.
L' is the lumped line inductance and C_{ser} is the series capacitor

3.13 Line Charging Current

Line charging current is the source current at no-load at the load end. For a line with the characteristic $ABCD$, we have:

$$\hat{V}_S = A \cdot \hat{V}_R, \quad \hat{I}_S = C \cdot \hat{V}_R \Rightarrow \hat{I}_{S\,charge} = \frac{C}{A} \hat{V}_S \tag{3.54}$$

$$\text{Short line:} \quad \hat{I}_{S\,charge} = 0 \tag{3.55}$$

$$\text{Medium line of } T \text{ model:} \quad \hat{I}_{S\,charge} = \frac{Y}{1+0.5ZY} \hat{V}_S \overset{**}{\simeq} Y \cdot \hat{V}_S \tag{3.56}$$

$$\text{Medium line of } \pi \text{ model:} \quad \hat{I}_{S\,charge} = \frac{Y(1+0.25ZY)}{1+0.5ZY} \hat{V}_S \overset{**}{\simeq} Y \cdot \hat{V}_S \tag{3.57}$$

$$\text{Long line:} \quad \hat{I}_{S\,charge} = \frac{\tanh(\gamma l)}{Z_C} \hat{V}_S \overset{**}{\simeq} Y \cdot \hat{V}_S \tag{3.58}$$

**: Note that $|Y| \ll |Z|$, so if Y tends to zero (in relation to Z), we find that $\hat{I}_{S\,charge} \overset{**}{\simeq} Y \cdot \hat{V}_S$ in relations (3.56)–(3.58).

3.14 Maximum Power Transfer

Assume:

$$\hat{V}_S = V_S \angle \delta, \quad \hat{V}_R = V_R \angle 0$$

$$T = \begin{bmatrix} A & B \\ C & D \end{bmatrix} = \begin{bmatrix} |A| \angle \varphi_A & |B| \angle \varphi_B \\ |C| \angle \varphi_C & |A| \angle \varphi_A \end{bmatrix}$$

$$P_R = \frac{V_S \cdot V_R}{|B|} \cos(\varphi_B - \delta) - \frac{|A| \cdot V_R^2}{|B|} \cos(\varphi_B - \varphi_A) \tag{3.59}$$

$$\Rightarrow P_{R\max} = P_R\big|_{\delta = \varphi_B} = \frac{V_S \cdot V_R}{|B|} - \frac{|A| \cdot V_R^2}{|B|} \cos(\varphi_B - \varphi_A) \tag{3.60}$$

We have the following relationship in the T model and long line exactly and in the π model and short line approximately:

$$\delta_{P_{R\max}} = \varphi_B = \tan^{-1}\left(\frac{X}{R}\right) \tag{3.61}$$

Part Two: Answer Question

3.15 Four-Choice Questions – 50 Questions

3.1. What is the A constant for a three-phase, transposed line, and
500 km, $f = 50\,\text{Hz}$, GMD $= 22\,\text{m}$, GMR $= 17\,\text{mm}$, $\cos\left(\dfrac{\pi}{9}\right) = 0.94$?

1. 0.87   2. 0.5   3. 0.9   4. 0.94

3.2. In a long line without losses with a frequency of 60 Hz, if a reactor with a size of Z_C is connected in the receiving end, what is the average length of the line in this case if $V_S = V_R$?

1. 1000   2. 1250   3. 1500   4. 2000

3.3. In a long line without losses with a frequency of 50 Hz, if a reactor with a size of 400 Ω is connected in the receiving end, what is the average length of the line in this case if $V_S = V_R$, $L = 16 \times 10^{-7}$ H/m and $C = 10^{-11}$ (F/m)?

1. 1000   2. 1250   3. 1042   4. 1500

3.4. If $X_{Lsh} = Z_C$, what is the maximum value of $\dfrac{V_S}{V_R}$ for a long line without losses?

1. 1   2. $\dfrac{\sqrt{2}}{2}$ 3. $\sqrt{2}$ 4. 2

3.5. In an asymmetric π model, we have: $\begin{bmatrix} A & B \\ C & D \end{bmatrix} = \begin{bmatrix} 1 & 4 \\ 1 & 5 \end{bmatrix}$.

What are Z, Y_R, and Y_S?

1. $Z=4$, $Y_R=0$, $Y_S=1$ 2. $Z=4$, $Y_R=0$, $Y_S=5/4$
3. $Z=3$, $Y_R=1$, $Y_S=1$ 4. $Z=3$, $Y_R=1$, $Y_S=5/4$

3.6. If Z_{OS} is the open-circuit impedance seen from the source terminal and Z_{OR} is the open-circuit impedance seen from the load terminal, what is $\dfrac{Z_{OS}}{Z_{OR}}$?

1. 1   2. $-\dfrac{A}{C}$ 3. -1 4. $\dfrac{A}{C}$

3.7. We put a capacitor at the end of a loss less transmission line so that the no-load voltage at the end of the line is equal to the voltage at the beginning. What is the impedance of this capacitor?

1. $\dfrac{\cos \beta l}{1 - \sin \beta l} Z_C$

2. $\dfrac{\sin \beta l}{\cos \beta l - 1} \dfrac{1}{Z_C}$

3. $\dfrac{\sin \beta l}{\cos \beta l - 1} Z_C$

4. $\dfrac{1 + \cos \beta l}{\sin \beta l} Z_C$

3.8. There are two lines with impedances and thermal capacitances.

$$I_{2\max} = 300\,\text{A}, \quad I_{1\max} = 200\,\text{A}, \quad Z_1 = 5\Omega\angle 30°, \quad Z_2 = 4\Omega\angle 30°$$

In parallel, these lines provide the load with a power factor of 0.8 lagging at 10 kV. What is the maximum load?

1. 500 A   2. 450 A   3. 400 A   4. 300 A

3.9. A 300 km transmission line has a shunt admittance of $6 \times 10^{-6}\ \mu\mho/\text{km}$. What is the transmission matrix of a shunt reactor that can compensate 50% of the total shunt admittance?

1. $\begin{bmatrix} 1 & -j0.0018 \\ 0 & 1 \end{bmatrix}$

2. $\begin{bmatrix} 1 & 0 \\ j0.0018 & 1 \end{bmatrix}$

3. $\begin{bmatrix} 1 & j0.0009 \\ 0 & 1 \end{bmatrix}$

4. $\begin{bmatrix} 1 & 0 \\ -j0.0009 & 1 \end{bmatrix}$

3.10. A load and a single-phase generator are connected by 20 km overhead transmission line. What is the percentage of voltage regulation?

$$\text{Load}: \quad 10\,\text{kV}, \ 5\,\text{MW}, \ 50\,\text{Hz}, \ 0.707\,\text{lag},$$

$$\text{Line}: \quad r = 0.02\,\Omega/\text{km}, \ x = 0.04\,\Omega/\text{km},$$

1. 4%   2. 5%   3. 6%   4. 7%

3.11. In question 3.10, where the word single-phase should be replaced with three-phase, what is the voltage regulation?

1. 18%   2. 6%   3. 2%   4. 7%

3.12. Assume a 1 km three-phase transmission line. The voltage at the source is 200 kV, and the parameters of the line are as follows. Calculate the maximum length of the line which will not exceed 250 kV at the end of the line in the open circuit.

$$r = 0.05\,\Omega/\text{km}, \quad b = 1\,\mu\mho/\text{km}, \quad x = 0.4\,\Omega/\text{km}$$

1. 2000 km   2. 1500 km   3. 1000 km   4. 1250 km

3.13. Calculate the "incident voltage wave" in the middle of a 200 kV transmission line that is lossless and no-load.

$$l = 400\,\text{km}, \quad x = 0.1\pi\,(\Omega/\text{km}), \quad y = j10\pi \times 10^{-6}\,(\text{℧}/\text{km})$$

1. $100k\cos(18°)\angle 18°$ 2. $100k\cos(36°)\angle 36°$

3. $100k\angle -36°$ 4. $100k\angle 36°$

3.14. What is the characteristic matrix of a 300 km long line when $z = j1\,(\Omega/\text{km})$, $y = j2\times10^{-6}\,(\text{℧}/\text{km})$?

1. $\begin{bmatrix} 0.9 & j707 \\ j600\mu & 0.9 \end{bmatrix}$ 2. $\begin{bmatrix} 0.91 & j291 \\ j582\mu & 0.91 \end{bmatrix}$

3. $\begin{bmatrix} 1 & j300 \\ 0 & 1 \end{bmatrix}$ 4. $\begin{bmatrix} 0.91 & j250 \\ j760\mu & 0.91 \end{bmatrix}$

3.15. What is the characteristic matrix of a 230 kV, 400 km long line when $r = 0.113\,\Omega/\text{km}$, $x = 0.61\,\Omega/\text{km}$, and $y = j3.2\times10^{-6}\,(\text{℧}/\text{km})$?

1. $\begin{bmatrix} 0.91\angle 15° & 214\angle 86° \\ 2.7\,\text{m}\angle 98° & 0.91\angle 15° \end{bmatrix}$ 2. $\begin{bmatrix} 0.85\angle 1.9° & 236\angle 80° \\ 1.2\,\text{m}\angle 90.5° & 0.85\angle 1.9° \end{bmatrix}$

3. $\begin{bmatrix} 0.92\angle 1.5° & 245\angle 85° \\ 1.5\mu\angle 84° & 0.92\angle 1.5° \end{bmatrix}$ 4. $\begin{bmatrix} 0.81\angle 5° & 1.5\angle 48° \\ 2.7\,\text{m}\angle 98° & 0.81\angle 5° \end{bmatrix}$

3.16. The two lines L_1 and L_2 with the following characteristic matrix are parallel to each other. What is A in the equivalent line characteristic matrix?

$$T_{L_1} = \begin{bmatrix} 0.9 & j200 \\ j2.761\,\text{m} & 0.9 \end{bmatrix} \quad \text{and} \quad T_{L_2} = \begin{bmatrix} 0.85 & j300 \\ j3.476\,\text{m} & 0.85 \end{bmatrix}$$

1. 0.86   2. 0.87   3. 0.88   4. 0.89

3.17. When n lines with the characteristic $A_1B_1C_1D_1$ are parallel, and then the side of the sender and receiver is changed, what is B in the characteristic matrix of the updated line?

1. $\dfrac{C_1}{n}$ 2. $-\dfrac{C_1}{n}$ 3. $\dfrac{B_1}{n}$ 4. $-\dfrac{B_1}{n}$

3.18. In a transmission line, the characteristic impedance is $Z_C = 1 - j$. With $Z_L = 1 + j$, the load is located at the end of the line. Can the reactive element be parallel to the load if the switch is connected at the beginning of the line to produce a zero-amplitude (reflected voltage wave)?

1. Inductor with reactance of 0.5

2. Capacitor with reactance of 0.5

3. Inductor with reactance of 1

4. Capacitor with reactance of 1

3.19. For the power system in Figure 3.6, the line is lossless and $\beta l = 30°$. What should be added to the busbar angle of 2 in order to make the active power across the line equal to one SIL?

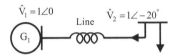

FIGURE 3.6
Question system 3.19.

1. 10°   2. −10°   3. 20°   4. −20°

3.20. $Z_{SS} = j2250\,(\Omega)$ and $Z_{SO} = -j4000\,(\Omega)$ are the short-circuit and the open-circuit impedance seen from the source terminal, respectively. What is A in the characteristic matrix?

1. 0.8   2. 0.85   3. 0.9   4. 0.95

3.21. If the line is modeled with a one-half of T model, what is A in the characteristic matrix for a three-phase, transposed 60 Hz, 200 km, 300 kV line?

$$z = j2\,(\Omega/\text{km}), \quad y = j2 \times 10^{-6}\,(\mho/\text{km})$$

1. 0.84   2. 0.92   3. 1.08   4. 0.96

3.22. For the medium line, we have the following characteristic matrix, about how much voltage regulation is present in this line.

$$V_{R,\,\text{no-load}} = 800\,\text{kV}, \quad V_S = 900\,\text{kV}, \quad \begin{bmatrix} 0.9\angle 0.2° & 80\angle 80° \\ 2 \times 10^{-3} \angle 95° & 0.9\angle 0.2° \end{bmatrix}$$

1. 12.5%   2. 25%   3. 20%   4. 1.25%

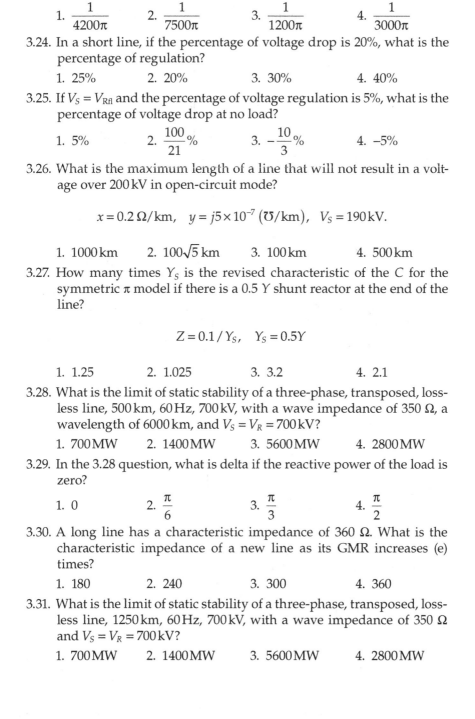

3.23. The series reactance of the 50 Hz short line is 30 Ω per phase. What is the capacitance in each phase in μF if we compensate with a series capacitor of 40%?

1. $\dfrac{1}{4200\pi}$ 2. $\dfrac{1}{7500\pi}$ 3. $\dfrac{1}{1200\pi}$ 4. $\dfrac{1}{3000\pi}$

3.24. In a short line, if the percentage of voltage drop is 20%, what is the percentage of regulation?

1. 25%   2. 20%   3. 30%   4. 40%

3.25. If $V_S = V_{Rfl}$ and the percentage of voltage regulation is 5%, what is the percentage of voltage drop at no load?

1. 5%   2. $\dfrac{100}{21}$%   3. $-\dfrac{10}{3}$%   4. –5%

3.26. What is the maximum length of a line that will not result in a voltage over 200 kV in open-circuit mode?

$$x = 0.2\,\Omega/\text{km}, \quad y = j5 \times 10^{-7}\,(\text{℧}/\text{km}), \quad V_S = 190\,\text{kV}.$$

1. 1000 km   2. $100\sqrt{5}$ km   3. 100 km   4. 500 km

3.27. How many times Y_S is the revised characteristic of the C for the symmetric π model if there is a 0.5 Y shunt reactor at the end of the line?

$$Z = 0.1/Y_S, \quad Y_S = 0.5Y$$

1. 1.25   2. 1.025   3. 3.2   4. 2.1

3.28. What is the limit of static stability of a three-phase, transposed, lossless line, 500 km, 60 Hz, 700 kV, with a wave impedance of 350 Ω, a wavelength of 6000 km, and $V_S = V_R = 700$ kV?

1. 700 MW   2. 1400 MW   3. 5600 MW   4. 2800 MW

3.29. In the 3.28 question, what is delta if the reactive power of the load is zero?

1. 0   2. $\dfrac{\pi}{6}$ 3. $\dfrac{\pi}{3}$ 4. $\dfrac{\pi}{2}$

3.30. A long line has a characteristic impedance of 360 Ω. What is the characteristic impedance of a new line as its GMR increases (e) times?

1. 180   2. 240   3. 300   4. 360

3.31. What is the limit of static stability of a three-phase, transposed, lossless line, 1250 km, 60 Hz, 700 kV, with a wave impedance of 350 Ω and $V_S = V_R = 700$ kV?

1. 700 MW   2. 1400 MW   3. 5600 MW   4. 2800 MW

3.32. In a long transmission line, we have $A = \sqrt{0.91}\angle 0$ and $B = j90$. What is the characteristic impedance?

    1. 300   2. 333   3. $j300$ 4. $j333$

3.33. In a 300 kV three-phase transposed line, V_S equals V_R and a shunt reactor is added at its end.

$$f = 50\,\text{Hz}, \quad \text{GMD} = 22\,\text{m}, \quad \text{GMR} = 17\,\text{mm}, \quad \cos\left(\frac{\pi}{5}\right) = 0.8, \quad l = 1000\,\text{km}$$

What is the voltage of a phase in the middle of the line?

    1. $50\,\text{kV}\sqrt{3}$ 2. $200\,\text{kV}\,\sqrt{3}$ 3. $200\,\text{kV}\dfrac{\sqrt{3}}{3}$ 4. $200\,\text{kV}$

3.34. In a transmission line, if $A = 0.8\angle 5°$, $B = 200\angle 80°$, what is C?

    1. $1.93\,\text{m}\angle 83.27°$ 2. $3.196\,\text{m}\angle -54.24°$

    3. $2.497\,\text{m}\angle 87.22°$ 4. $8.170m\angle 3.9°$

3.35. In a long line, when $Y = j2.7 \times 10^{-3}\,\mho$ and $Z_C \cdot \gamma l = 120\angle 75°$, what is γl magnitude?

    1. $\dfrac{2}{3}\sqrt{10}$ 2. $\dfrac{20}{3}\sqrt{10}$ 3. $0.18\sqrt{10}$ 4. 0.324

3.36. Suppose Z_{sc} and Z_{so} are the short-circuit impedance and open-circuit impedance seen at the source, respectively. Which of the following equations is true for characteristic impedance?

    1. $Z_C = Z_{OC} - Z_{SC}$ 2. $Z_C = Z_{SC} - Z_{OC}$

    3. $Z_C = \sqrt{Z_{OC} \times Z_{SC}}$ 4. $Z_C = \dfrac{1}{2}(Z_{OC} + Z_{SC})$

3.37. The *ABCD* constants of the two systems are:

$$\text{System 1:} \begin{bmatrix} A_1 & B_1 \\ C_1 & D_1 \end{bmatrix} = \begin{bmatrix} 1 & 2j \\ 0 & 1 \end{bmatrix}$$

$$\text{System 2:} \begin{bmatrix} A_2 & B_2 \\ C_2 & D_2 \end{bmatrix} = \begin{bmatrix} 1 & 0 \\ 2j & 1 \end{bmatrix}$$

Let us assume that the two systems above are cascaded, and that system 1 is the input and system 2 is the output. What is the *ABCD* constant of the equal system?

    1. $\begin{bmatrix} -3 & 2j \\ 2j & 1 \end{bmatrix}$ 2. $\begin{bmatrix} 5 & 2j \\ 2j & 1 \end{bmatrix}$

    3. $\begin{bmatrix} 2 & 2j \\ 2j & 2 \end{bmatrix}$ 4. None of them

3.38. A three-phase symmetric transmission line with a length of 150 km is assumed to be the T model. What is the line charging current if the series impedance of the line is Z and its capacitive admittance is Y?

1. $I_{S \text{ charge}} = ZV_R(1+0.5ZY)^{-1}$ 2. $I_{S \text{ charge}} = YV_S(1+0.5YZ)^{-1}$

3. $I_{S \text{ charge}} = ZV_S(1+0.5YZ)^{-1}$ 4. $I_{S \text{ charge}} = YV_R(1+0.5YZ)^{-1}$

3.39. What is the Ferranti effect and when does it occur on a long transmission line?

1. Voltage increases at the receiving end when the load is low

2. Voltage increases at the receiving end when the load is full

3. Voltage increases at the receiving end in short-circuit condition

4. Voltage decreases at the receiving end when the load is low

3.40. Three-phase transmission lines without losses have $ABCD$ constants. What are the updated $ABCD$ constants if a series capacitor is added to both sides of each phase in the transmission line with a reactance of $0.5x_c$?

1. $\begin{bmatrix} \dfrac{-jx_c}{2} & 1 \\ 0 & 1 \end{bmatrix} \begin{bmatrix} A & B \\ C & D \end{bmatrix} \begin{bmatrix} \dfrac{-jx_c}{2} & 1 \\ 0 & 1 \end{bmatrix}$

2. $\begin{bmatrix} 0 & 1 \\ \dfrac{-jx_c}{2} & 1 \end{bmatrix} \begin{bmatrix} A & B \\ C & D \end{bmatrix} \begin{bmatrix} 0 & 1 \\ \dfrac{-jx_c}{2} & 1 \end{bmatrix}$

3. $\begin{bmatrix} 1 & \dfrac{-jx_c}{2} \\ 0 & 1 \end{bmatrix} \begin{bmatrix} A & B \\ C & D \end{bmatrix} \begin{bmatrix} 1 & \dfrac{-jx_c}{2} \\ 0 & 1 \end{bmatrix}$

4. $\begin{bmatrix} 0 & 1 \\ 1 & \dfrac{-jx_c}{2} \end{bmatrix} \begin{bmatrix} A & B \\ C & D \end{bmatrix} \begin{bmatrix} 0 & 1 \\ 1 & \dfrac{-jx_c}{2} \end{bmatrix}$

3.41. An area is supplied by a short line that has a single-phase model as shown below. There is a constant capacitor in this area of the bus. Which of the following relationships is true (Figure 3.7)?

FIGURE 3.7
Question system 3.41.

1.
$$\begin{bmatrix} V_S \\ I_S \end{bmatrix} = \begin{bmatrix} 1+ZY & 1 \\ Z & Y \end{bmatrix} \begin{bmatrix} V_R \\ I_R \end{bmatrix}$$

2.
$$\begin{bmatrix} V_S \\ I_S \end{bmatrix} = \begin{bmatrix} 1+ZY & Z \\ Y & 1 \end{bmatrix} \begin{bmatrix} V_R \\ I_R \end{bmatrix}$$

3.
$$\begin{bmatrix} V_S \\ I_S \end{bmatrix} = \begin{bmatrix} 1+ZY & 1 \\ Y & Z \end{bmatrix} \begin{bmatrix} V_R \\ I_R \end{bmatrix}$$

4.
$$\begin{bmatrix} V_S \\ I_S \end{bmatrix} = \begin{bmatrix} 1+ZY & Y \\ Z & 1 \end{bmatrix} \begin{bmatrix} V_R \\ I_R \end{bmatrix}$$

3.42. What are the A and B coefficients in the transmission matrix in Figure 3.8?

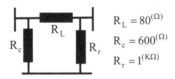

$R_L = 80^{(\Omega)}$

$R_c = 600^{(\Omega)}$

$R_r = 1^{(K\Omega)}$

FIGURE 3.8
Question system 3.42.

1. $A = 1.07\ \Omega,\ B = 90\ \Omega$ 2. $A = 1.07\ \Omega,\ B = 80\ \Omega$

3. $A = 1.08\ \Omega,\ B = 90\ \Omega$ 4. $A = 1.08\ \Omega,\ B = 80\ \Omega$

3.43. In the following scenario, an area is supplied by a short three-phase transmission line, whose single-phase model appears in the following. If we connect a resistance load (R) in the middle of the line, which one of the following relationships is correct (Figure 3.9)?

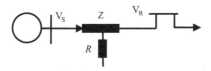

FIGURE 3.9
Question system 3.43.

1.
$$\begin{bmatrix} V_S \\ I_S \end{bmatrix} = \begin{bmatrix} R+Z & Z(R+Z) \\ Z & R+Z \end{bmatrix} \begin{bmatrix} V_R \\ I_R \end{bmatrix}$$

2.
$$\begin{bmatrix} V_S \\ I_S \end{bmatrix} = \begin{bmatrix} 1+Z & Z(R+Z) \\ Z & R+Z \end{bmatrix} \begin{bmatrix} V_R \\ I_R \end{bmatrix}$$

3.
$$\begin{bmatrix} V_S \\ I_S \end{bmatrix} = \begin{bmatrix} 1+\dfrac{Z}{2R} & Z+\dfrac{Z^2}{4R} \\ \dfrac{1}{R} & 1+\dfrac{Z}{2R} \end{bmatrix} \begin{bmatrix} V_R \\ I_R \end{bmatrix}$$

4.
$$\begin{bmatrix} V_S \\ I_S \end{bmatrix} = \begin{bmatrix} 1+\dfrac{Z}{2R} & Z+\dfrac{Z^2}{2R} \\ \dfrac{1}{R} & 1+\dfrac{Z}{4R} \end{bmatrix} \begin{bmatrix} V_R \\ I_R \end{bmatrix}$$

3.44. The $ABCD$ constants in the single-phase model of a three-phase transmission line are as follows. The voltage angle at the end of the line is zero (reference) and the voltage angle at the beginning of the line is delta. When is the maximum active power at the end of the line?

$$\begin{bmatrix} |A|\angle\alpha & |B|\angle\beta \\ |C|\angle\theta & |A|\angle\alpha \end{bmatrix}$$

1. $\delta = \beta$ 2. $\delta = \alpha$ 3. $\alpha = \beta$ 4. $\alpha = \theta$

3.45. An open-circuit impedance seen at the source of a long transmission line is the inverse of the short-circuit impedance seen at the source. Which of the following relationships is true?

1. $A + B = \dfrac{1}{A - B}$ 2. $A + B = \dfrac{1}{B - A}$

3. $A = \sqrt{1 - B^2}$ 4. $A = \sqrt{B^2 - 1}$

3.46. A series capacitor with a constant of B_C is placed in the middle of a transmission line whose two parts have constants of $ABCD$. What is the equivalent B constant for the transmission line as a whole?

1. $A^2 + C(AB_C + B)$ 2. $BC + D(CB_C + D)$

3. $AB + D(AB_C + B)$ 4. $AC + C(CB_C + D)$

3.47. The SIL of a lossless transmission line of 400 kV will equal 100 MW. What will be the capacitance for each phase of the line in (F/m)?

1. $\dfrac{10^{-10}}{144}$ 2. $\dfrac{10^{-10}}{48}$ 3. $\dfrac{10^{-10}}{24}$ 4. $\dfrac{10^{-10}}{16}$

3.48. Each phase of a line has an equivalent circuit of 200 kV as shown in Figure 3.10. Assume the base power is 100 MVA. When the voltage of the source is 1 p.u., the load power is zero ($S_R = P_R + jQ_R = 0$), and we want the load voltage to also be 1 p.u., what is the shunt compensating impedance at the end of the line?

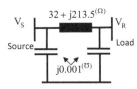

FIGURE 3.10
Question system 3.48.

1. $j0.4$ p.u.   2. $-j0.4$ p.u.   3. $j2.5$ p.u.   4. $-j2.5$ p.u.

3.49. In the Figure 3.11, a lossless transmission line with a characteristic impedance of Z_C, an ideal generator with the effective voltage of $|V|$ and R, and a pure resistive load with SIL power are assumed. When the k breaker is closed, what is the maximum voltage of the load?

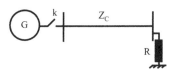

FIGURE 3.11
Question system 3.49.

1. $|V|$ 2. $\sqrt{2}|V|$ 3. $2\sqrt{2}|V|$
4. It cannot be calculated with the given information.

3.50. For a three-phase overhead transmission line of medium length and the π model, $\dfrac{Y}{2} = j\dfrac{B}{2}$ and $Z = R + jX$. If we use the short line model to calculate the voltage at the beginning of this line (V_S) with the known voltage at the end of the line (V_R), what is the calculation error compared to using the π model?

1. $|\Delta V_S| = \dfrac{BR}{\sqrt{2}}|V_R|$ 2. $|\Delta V_S| = \dfrac{BX}{\sqrt{2}}|V_R|$

3. $|\Delta V_S| = \dfrac{B\sqrt{R^2 + X^2}}{2}|V_R|$ 4. $|\Delta V_S| = \dfrac{\sqrt{R^2 + X^2}}{B\sqrt{2}}|V_R|$

3.16 Key Answers to Four-Choice Questions

Question	1	2	3	4
1. (1)	×			
2. (2)		×		
3. (2)		×		
4. (3)			×	
5. (1)	×			
6. (3)			×	
7. (3)			×	
8. (2)		×		
9. (4)				×

(Continued)

Question	1	2	3	4
10. (3)			×	
11. (2)		×		
12. (3)			×	
13. (4)				×
14. (2)		×		
15. (2)		×		
16. (3)			×	
17. (4)				×
18. (4)				×
19. (2)		×		
20. (1)	×			
21. (1)	×			
22. (2)		×		
23. (3)			×	
24. (1)	×			
25. (4)				×
26. (1)	×			
27. (3)			×	
28. (4)				×
29. (2)		×		
30. (3)			×	
31. (2)		×		
32. (1)	×			
33. (4)				×
34. (1)	×			
35. (3)			×	
36. (3)			×	
37. (1)	×			
38. (2)		×		
39. (1)	×			
40. (3)			×	
41. (2)		×		
42. (3)				×
43. (3)			×	
44. (1)	×			
45. (1)	×			
46. (3)			×	
47. (2)		×		
48. (3)			×	
49. (2)		×		
50. (3)			×	

3.17 Descriptive Answers to Four-Choice Questions

3.1. **Option 1 is correct.** Since GMD and GMR are the same for L and C, the relation (equation 3.36) can be used.

$$\beta l = \frac{\omega}{3} \times 10^{-8} \times l = \frac{2\pi \times 50}{3} \times 10^{-8} \times 500 \times 10^3 = \frac{\pi}{6}$$

$$\Rightarrow A = \cosh(\gamma l) = \cos(\beta l) = \cos\left(\frac{\pi}{6}\right) = \frac{\sqrt{3}}{2} = \frac{1.73}{2} = 0.87$$

3.2. **Option 2 is correct.** Using equation (3.48) we have:
It is said in the question: $X_{Lsh} = Z_C$

$$X_{Lsh} = \frac{\sin(\beta l)}{1 - \cos(\beta l)} \cdot Z_C \Rightarrow 1 = \frac{\sin(\beta l)}{1 - \cos(\beta l)} \Rightarrow \begin{cases} \sin(\beta l) = 1 - \cos(\beta l) \\ (\sin(\beta l))^2 + (\cos(\beta l))^2 = 1 \end{cases}$$

$$\Rightarrow \begin{cases} \cos(\beta l) = 0 \\ \sin(\beta l) = 1 \end{cases} \Rightarrow \beta l = \frac{\pi}{2} \stackrel{(3.36)}{\Rightarrow} \frac{2\pi \times 60}{3} \times 10^{-8} \times l = \frac{\pi}{2} \Rightarrow l = 1250\,\text{km}$$

The information from L and C is not given, but can be used from the approximate relation of equation (3.36).

3.3. **Option 2 is correct.** Considering that C and L are given in this question, they cannot be used from the approximate relationship of equation (3.36) but from its exact relationship, which is equation (3.34). Using relationships (3.35) and (3.48) we get:

$$Z_C = \sqrt{\frac{L}{C}} = \sqrt{\frac{16 \times 10^{-7}}{10^{-11}}} = 4 \times 10^2 = 400\,\Omega$$

$$\Rightarrow X_{Lsh} = \frac{\sin(\beta l)}{1 - \cos(\beta l)} \cdot Z_C \Rightarrow 1 = \frac{\sin(\beta l)}{1 - \cos(\beta l)} \Rightarrow \begin{cases} \sin(\beta l) = 1 - \cos(\beta l) \\ (\sin(\beta l))^2 + (\cos(\beta l))^2 = 1 \end{cases}$$

$$\Rightarrow \begin{cases} \cos(\beta l) = 0 \\ \sin(\beta l) = 1 \end{cases} \Rightarrow \beta l = \frac{\pi}{2} \stackrel{(3.34)}{\Rightarrow} 2\pi \times 50\sqrt{16 \times 10^{-7} \times 10^{-11}} \times l = \frac{\pi}{2}$$

$$\Rightarrow 2\pi \times 50 \times 4 \times 10^{-9} \times l = \frac{\pi}{2} \Rightarrow l = 1250\,\text{km}$$

3.4. **Option 3 is correct.** By using equation (3.47) we have:

$$X_{Lsh} = \frac{\sin(\beta l)}{\frac{V_S}{V_R} - \cos(\beta l)} \cdot Z_C \Rightarrow \sin(\beta l) = \frac{V_S}{V_R} - \cos(\beta l) \Rightarrow \frac{V_S}{V_R} = \sin(\beta l) + \cos(\beta l)$$

$$\Rightarrow \max\left(\frac{V_S}{V_R}\right) = \max\left(\sin(\beta l) + \cos(\beta l)\right)^{\beta l = \frac{\pi}{4}} = \sqrt{2}$$

3.5. **Option 1 is correct.** Comparing the asymmetric π model with equation (3.30), we get: $Z=4$, $Y_S=1$, $Y_R=0$

3.6. **Option 3 is correct.** Using equations (3.1) and (3.2) we have:

$$\left.\begin{array}{l} Z_{OS} = \left.\dfrac{\hat{V}_S}{\hat{I}_S}\right|_{\hat{I}_R=0} = \dfrac{A \cdot \hat{V}_R}{C \cdot \hat{V}_R} = \dfrac{A}{C} \\[3mm] Z_{OR} = \left.\dfrac{\hat{V}_R}{\hat{I}_R}\right|_{\hat{I}_S=0} = \dfrac{D \cdot \hat{V}_S}{-C \cdot \hat{V}_S} = -\dfrac{A}{C} \end{array}\right\} \Rightarrow \dfrac{Z_{OS}}{Z_{OR}} = -1$$

3.7. **Option 3 is correct.** Using equation (3.48), X_{Lsh} converts to $-X_C$ and then calculates X_C.

3.8. **Option 2 is correct.** We have: $\dfrac{\hat{I}_1}{\hat{I}_2} = \dfrac{Z_2}{Z_1}$

At first we assume $I_2 = I_{2\,max} = 300$, so we have:

$$\Rightarrow I_1 = \left|\frac{Z_2}{Z_1}\right| I_2 = \left|\frac{4\angle 30}{5\angle 30}\right| \times 300 = 240 > (200 = I_{1\,max})$$

So should $I_1 = I_{1\,max} = 200$. We have:

$$\Rightarrow I_2 = \left|\frac{Z_1}{Z_2}\right| I_1 = \left|\frac{5\angle 30}{4\angle 30}\right| \times 200 = 250 < (300 = I_{2\,max})$$

When Z_1 and Z_2 are in the same phase, then $\hat{I}_1$ and $\hat{I}_2$ are also in the same phase, so the total current equals:

$$I_t = I_{1\,max} + I_2 = 200 + 250 = 450\,A$$

3.9. **Option 4 is correct.** Using equation (3.50) we have:

Reactor admittance required: $B = \dfrac{50}{100} \times 6 \times 10^{-6} \times 300 = 0.0009\,\mho$

And using equation (3.21) with $Y = -jB = -j0.0009$, option 4 is correct.

3.10. **Option 3 is correct.** Based on the data, it appears that the desired line should be short.

$$V_{Rfl} = V_R = 10\,\text{kV}, \quad V_{Rnl} = \frac{V_S}{A} = \frac{V_S}{1} = V_S$$

$$R = 0.02 \times 20 = 0.4\,\Omega, \quad X = 0.04 \times 20 = 0.8\,\Omega$$

$$R\% = \frac{V_S - V_R}{V_R} \times 100$$

1-Exact method:

$$P = VI\cos(\varphi) \Rightarrow I_R = \frac{5\,\text{MW}}{10\,\text{kV} \times 0.707} = 707\,\text{A}$$

$$\Rightarrow \hat{I}_R = 707\angle - \cos^{-1}0.707 = 707\angle - 45°$$

$$\Rightarrow \hat{V}_S = (0.4 + j0.8)(707\angle - 45°) + 10\,\text{kV} = 10.602\,\text{kV}\angle 1.081°$$

$$\Rightarrow R\% = \frac{10.602\,\text{kV} - 10\,\text{kV}}{10\,\text{kV}} \times 100 = 6.018\%$$

We usually use the approximate method, which is faster than the exact method, since it doesn't require a calculator. However, accurate voltage regulation is more than just approximate voltage regulation.

2-Approximate method using equation (3.24):

$$R\% = \frac{R \cdot P_R + X \cdot Q_R}{V_R^2} \times 100 = \frac{0.4 \times 5\,\text{MW} + 0.8 \times 5\,\text{MW} \times \tan(45°)}{(10\,\text{kV})^2} \times 100$$

$$= \frac{(0.4 + 0.8) \times 5\,\text{M}}{100\,\text{M}} \times 100 = 6\%$$

3.11. **Option 2 is correct.** The percentage of voltage regulation does not change. In equation (3.24) and in a three-phase system, if P and Q are three-phase, V_R must be the line voltage, while if P and Q are single phase, V_R must be the single-phase voltage.

3.12. **Option 3 is correct.** When the line is no-load or open-circuit, we have:

$$\begin{bmatrix} \hat{V}_S \\ \hat{I}_S \end{bmatrix} = \begin{bmatrix} A & B \\ C & D \end{bmatrix} \begin{bmatrix} \hat{V}_R \\ 0 \end{bmatrix} \Rightarrow |A| = \frac{V_S}{V_R} = \frac{200}{250} = 0.8$$

Because the line resistance is much smaller than the line reactance, it can be ignored.

$$A = 1 + \frac{ZY}{2} = 1 + \frac{(0.05l + j0.4l) \times j \times 10^{-6} \times l}{2} \simeq 1 + \frac{(j0.4) \times j \times 10^{-6} \times l^2}{2}$$

$$\Rightarrow A = 1 - 0.2 \times 10^{-6} \times l^2 = 0.8 \Rightarrow l = 1000 \, \text{km}$$

3.13. **Option 4 is correct.** Using equations (3.32) and (3.33) we have: (No-load: $I_R = 0$)

Incident voltage wave (Send):

$$\hat{V}_{send} = \frac{V_R}{2} \sqrt{3} \times e^{\alpha x} \times e^{j\beta x} = \frac{V_R}{2} \sqrt{3} \times e^{\alpha x} \angle \beta x$$

(x) is distance from the beginning of the line and V_R is the voltage of the load phase.

$$x = \frac{l}{2} = 200 \, \text{km}$$

$$\gamma = \sqrt{z \cdot y} = \sqrt{j0.1\pi \times j10\pi \times 10^{-6}} = 0 + j\pi \times 10^{-3} = \alpha + j\beta$$

$$\Rightarrow \alpha = 0, \ \beta = \pi \times 10^{-3} \, (\text{rad/km}) \Rightarrow \beta = \pi \times 10^{-3} \times \frac{180}{\pi} = 0.18° \, (1/\text{km})$$

$$\Rightarrow \hat{V}_{send} = \frac{200 \, \text{kV}}{2 \times \sqrt{3}} \times \sqrt{3} \times e^0 \angle (0.18 \times 200) = 100 \, \text{kV} \angle 36°$$

3.14. **Option 2 is correct.** If you want to solve the problem without the use of a calculator, you can use the long-line approximation method or by using equation (3.14).

$$Z = j1 \times 300 = j300 \, (\Omega), \quad Y = j600 \, (\mu\text{U})$$

$$A = D = 1 + \frac{ZY}{2} = 1 - \frac{300 \times 600 \, \mu}{2} = 1 - \frac{0.18}{2} = 0.91$$

$$B = Z\left(1+\frac{ZY}{6}\right) = j300\left(1-\frac{0.18}{6}\right) = j300 \times 0.97 = j291$$

$$C = Y\left(1+\frac{ZY}{6}\right) = j600\mu(0.97) = j582\,\mu$$

Quick solution: This is not a short line model, so option 3 is incorrect. B is certainly less than Z, so option 1 is also incorrect, but B is not too far from Z, so option 4 is also wrong.

3.15. **Option 2 is correct.** The method of quick solving must be applied to this problem. We can determine the correct answer if we pay attention to the π model.

Option 1 is incorrect since A has a large angle. (A) must be close to $1\angle 0°$.

Option 3 is incorrect because the angle of C is smaller than 90°.
Option 4 is incorrect because B is low. (B) must be close to Z.

3.16. **Option 3 is correct.** Using equation (3.16) we have:

$$A = \frac{A_1 \cdot B_2 + B_1 \cdot A_2}{B_1 + B_2} = \frac{0.9 \times j300 + j200 \times 0.85}{j500} = \frac{0.9 \times 3 + 2 \times 0.85}{5} = \frac{4.4}{5} = 0.88$$

3.17. **Option 4 is correct.** Option 4 is obtained by combining relations (equation 3.2) and (equation 3.18).

3.18. **Option 4 is correct.** From the relation of the reflected voltage wave (equation 3.33), we have:

$$A_2 = \left|\frac{\hat{V}_R - Z_C \hat{I}_R}{2}\right| = 0 \Rightarrow \hat{V}_R = Z_C \hat{I}_R \Rightarrow \frac{\hat{V}_R}{\hat{I}_R} = Z_C = 1 - j$$

$$\frac{\hat{V}_R}{\hat{I}_R} = z\|(1+j) = \frac{(1+j)z}{1+j+z} = 1 - j \Rightarrow z + jz = (1+j+z) - j + 1 - jz$$

$$2jz = 2 \Rightarrow z = -j$$

3.19. **Option 2 is correct.** Using equation (3.46), we have:

$$V_S = V_R = 1\,\text{p.u.}$$

$$P_{3ph} = \frac{V_S(\text{p.u.}) \cdot V_R(\text{p.u.}) \cdot \text{SIL}}{\sin(\beta l)} \sin(\delta) = \text{SIL} \Rightarrow \sin(\beta l) = \sin(\delta)$$

$$\Rightarrow \beta l = \delta \Rightarrow 30 = 0 - (-20 + x) \Rightarrow x = -10$$

3.20. **Option 1 is correct.** Using equation (3.27) we have:

$$A = \sqrt{\frac{Z_{SO}}{Z_{SO} - Z_{SS}}} = \sqrt{\frac{-j4000}{-j4000 - j2250}} = \sqrt{\frac{4000}{6250}} = \sqrt{\frac{4 \times 100}{25 \times 25}} = \sqrt{\frac{4 \times 4}{25}}$$

$$= \sqrt{\frac{16 \times 4}{100}} = \sqrt{0.64} = 0.8$$

3.21. **Option 1 is correct.** Using equation (3.29), we have: $A = 1 + ZY$

$$Z = j2 \times 200 = j400\,(\Omega), \quad Y = j2 \times 10^{-6} \times 200 = j400\,(\mu\text{℧})$$

$$\Rightarrow A = 1 + j400 \times j400\mu = 1 - 0.16 = 0.84$$

3.22. **Option 2 is correct.** Using equations (3.22) or (3.25), we have:

$$V_{Rfl} = 800\,\text{kV}, \quad V_{Rnl} = \frac{V_S}{|A|} = \frac{900\,\text{kV}}{0.9} = 1000\,\text{kV}$$

$$R\% = \frac{1000 - 800}{800} \times 100 = 25\%$$

3.23. **Option 3 is correct.**

Capacitance reactance required: $X_C = \dfrac{40}{100} \times 30 = 12\,\Omega/\text{ph}$

$$12 = \frac{1}{\omega C} \Rightarrow C = \frac{1}{2\pi \times 50 \times 12} = \frac{1}{1200\pi}$$

3.24. **Option 1 is correct.** Using equations (3.25) and (3.26), and this issue, which in the short line is $A = 1$, we have:

$$\left.\begin{aligned}
\text{Reg} &= \frac{V_S - V_R}{V_R} = \frac{V_S}{V_R} - 1 \\[2mm]
\Delta V &= \frac{V_S - V_R}{V_S} = 1 - \frac{V_R}{V_S} \Rightarrow \frac{V_R}{V_S} = 1 - \Delta V
\end{aligned}\right\} \Rightarrow \text{Reg} = \frac{1}{1 - \Delta V} - 1 = \frac{\Delta V}{1 - \Delta V}$$

$$\Rightarrow \text{Reg} = \frac{0.2}{1 - 0.2} = 0.25 \Rightarrow \text{Reg}\% = 25\%$$

3.25. **Option 4 is correct.** Using equations (3.22) or (2.25), we have:

$$R\% = \frac{V_{Rnl} - V_{Rfl}}{V_{Rfl}} \times 100, \quad V_{Rnl} = \frac{V_S}{A}, \quad V_S = V_{Rfl} \triangleq V_R$$

$$\Rightarrow R\% = \frac{\frac{V_R}{A} - V_R}{V_R} \times 100 = \frac{1-A}{A} \times 100$$

Also by using equation (3.26) for the percentage of voltage drop in the no-load mode we have:

$$\Delta V\% = \frac{V_S - V_R}{V_S} \times 100,$$

$$V_S = A \cdot V_R, \quad \Rightarrow \Delta V\% = \frac{A \cdot V_R - V_R}{A \cdot V_R} \times 100 = \frac{A-1}{A} \times 100$$

Therefore, it can be seen in this question: $\Delta V\% = -R\% = -5\%$

3.26. **Option 1 is correct.** In open-circuit mode, we have: (similar to 3.12 except without resistance)

$$\begin{bmatrix} \hat{V}_s \\ \hat{I}_s \end{bmatrix} = \begin{bmatrix} A & B \\ C & D \end{bmatrix} \begin{bmatrix} \hat{V}_R \\ 0 \end{bmatrix} \Rightarrow |A| = \frac{V_S}{V_R} = \frac{190}{200} = 0.95$$

$$A = 1 + \frac{ZY}{2} = 1 - \frac{xb}{2} \times l^2 = 1 - 0.5 \times 0.2 \times 5 \times 10^{-7} \times l^2$$

$$= 1 - 5 \times 10^{-8} l^2 = 0.95 \Rightarrow l^2 = \frac{0.05}{5 \times 10^{-8}} = 10^6 \Rightarrow l = 1000 \, km$$

3.27. **Option 3 is correct.** Using equation (3.30), we have:

$$Y_S = \frac{Y}{2}, \quad Y_R = \frac{Y}{2} + Y_S = 2Y_S,$$

$$Z = \frac{0.1}{Y_S}, \quad \Rightarrow C = Y_S + 2Y_S + Y_S \cdot 2Y_S \cdot \frac{0.1}{Y_S} = 3.2Y_S$$

3.28. **Option 4 is correct.** Using equation (3.44) with $\sin \delta = 1$, we have:

Static stability limit: $P_{max} = \dfrac{V_S \times V_R}{Z_C \cdot \sin\left(\dfrac{2\pi l}{\lambda}\right)} = \dfrac{700\,kV \times 700\,kV}{350 \cdot \sin\left(\dfrac{2\pi \times 500}{6000}\right)} = 2800\,MW$

3.29. **Option 2 is correct.** Using equation (3.45) with $V_S = V_R$, we have:

$$Q_{R3ph} = \frac{V_S^2}{X'}\left(\cos(\delta) - \cos(\beta l)\right) = 0 \Rightarrow \delta = \beta l = \frac{2\pi l}{\lambda} = \frac{2\pi \times 500}{6000} = \frac{\pi}{6}$$

3.30. **Option 3 is correct.** Using equation (3.35), we have:

$$Z_{C1} = 360 = 60\ln\left(\frac{GMD}{GMR_1}\right) \Rightarrow \frac{GMD}{GMR_1} = e^6, \ GMR_2 = e \times GMR_1$$

$$\Rightarrow Z_{C2} = 60\ln\left(\frac{GMD}{GMR_2}\right) = 60\ln\left(\frac{GMR_1 \times e^6}{e \times GMR_1}\right) = 60\ln\left(e^5\right) = 300$$

3.31. **Option 2 is correct.** The problem is similar to question **3.28** except that lambda is not given, so lambda must be calculated using an approximate relation. Using equation (3.38), we have:

$$\lambda = \frac{3 \times 10^8}{60} = 5 \times 10^6 \, m = 5000 \, km$$

Using equation (3.44) with $\sin \delta = 1$, we have:

$$P_{max} = \frac{V_S \times V_R}{Z_C \cdot \sin\left(\frac{2\pi l}{\lambda}\right)} = \frac{700\,kV \times 700\,kV}{350 \cdot \sin\left(\frac{2\pi \times 1250}{5000}\right)} = \frac{2\,kV \times 700\,kV}{\sin\left(\frac{\pi}{2}\right)} = 1400\,MW$$

3.32. **Option 1 is correct.** Using the long line model. we have:

$$Z_C = \sqrt{\frac{B}{C}}, \ A^2 - BC = 1$$

$$\Rightarrow C = \frac{A^2 - 1}{B} \Rightarrow Z_C = \frac{B}{\sqrt{A^2 - 1}} = \frac{j90}{\sqrt{0.91 - 1}} = \frac{j90}{j0.3} = 300$$

3.33. **Option 4 is correct.** The problem is similar to question **3.1**. Since GMD and GMR are the same for L and C, the relation of equation (3.36) can be used.

$$\beta l = \frac{\omega}{3} \times 10^{-8} \times l = \frac{2\pi \times 50}{3} \times 10^{-8} \times 1000 \times 10^3 = \frac{\pi}{3}$$

Using equation (3.49), we have (for single phase):

$$V_m = \frac{V_R}{\cos\left(\frac{\beta l}{2}\right)} = \frac{300\,kV}{\sqrt{3} \times \cos\left(\frac{\pi}{6}\right)} = \frac{300\,kV}{\sqrt{3} \times \frac{\sqrt{3}}{2}} = 200\,kV$$

3.34. **Option 1 is correct**. Since the problem should not be solved with a calculator, experience and approximation are used. C must be close to $0\angle90°$ (similar $y = jb$) so answers 2 and 4 are incorrect. Between options 1 and 3, we need to determine the approximate value of C.

$$A^2 - BC = 1 \Rightarrow C = \frac{A^2 - 1}{B} \Rightarrow |C| = \left|\frac{0.8^2 - 1}{200}\right| = \left|\frac{0.64 - 1}{200}\right| = \frac{0.36}{200} = 1.8 \text{ m}$$

So option 1 is correct.

3.35. **Option 3 is correct**. Using equations (3.9) and (3.10), we have:

$$Z_C \cdot \gamma l = Z, \quad \gamma l = \sqrt{ZY}$$

$$|\gamma l| = \sqrt{120 \times 2.7 \times 10^{-3}} = \sqrt{4 \times 30 \times 3 \times 0.9 \times 10^{-3}} = \frac{2 \times 3 \times 3}{10\sqrt{10}} = 0.18\sqrt{10}$$

3.36. **Option 3 is correct**. As a result of equation (3.27), we can prove it.

3.37. **Option 1 is correct**.

$$T = T_1 \times T_2 = \begin{bmatrix} 1 & 2j \\ 0 & 1 \end{bmatrix} \begin{bmatrix} 1 & 0 \\ 2j & 1 \end{bmatrix} = \begin{bmatrix} -3 & 2j \\ 2j & 1 \end{bmatrix}$$

3.38. **Option 2 is correct**. Line charging current is the source current at no-load at the load end. Using equation (3.56), Option 2 is correct.

3.39. **Option 1 is correct**.

3.40. **Option 3 is correct**. The series capacitor model is similar to the short-line model.

$$T_c = \begin{bmatrix} 1 & \dfrac{-jx_c}{2} \\ 0 & 1 \end{bmatrix} \Rightarrow T' = T_c \cdot T \cdot T_c$$

3.41. **Option 2 is correct**. When a Y admittance is connected to the end of the short line, it becomes one-half of the T model, so option 2 is correct using equation (3.29).

3.42. **Option 4 is correct**. From the asymmetric π model (relation of equation 3.30), we have:

$$A = 1 + ZY_R = 1 + R_L \times \frac{1}{R_r} = 1 + 80 \times \frac{1}{1000} = 1.08, \quad B = Z = R_L = 80$$

3.43. **Option 3 is correct**. The line model is similar to the T model, in which the series impedance is Z and the parallel admittance is $Y = \dfrac{1}{R}$.

The second solution: If $R = \infty$, it should look like the short line model, so options 3 or 4 are correct, so it should be $A = D$, so option 3 is correct.

3.44. **Option 1 is correct**. Using (equation 3.61), option 1 is correct.

3.45. **Option 1 is correct**. Using (equation 3.27), we have:

$$Z_{SO} = \frac{1}{Z_{SS}} \Rightarrow \frac{A}{C} = \frac{A}{B} \Rightarrow B = C, \quad A^2 - BC = 1 \Rightarrow A^2 - B^2 = 1$$

3.46. **Option 3 is correct**. The series capacitor characteristic matrix is similar to a short line. We can multiply the matrix as follows:

$$\begin{bmatrix} A & B \\ C & D \end{bmatrix} \begin{bmatrix} 1 & B_C \\ 0 & 1 \end{bmatrix} \begin{bmatrix} A & B \\ C & D \end{bmatrix} = \begin{bmatrix} A & AB_C + B \\ C & CB_C + D \end{bmatrix} \begin{bmatrix} A & B \\ C & D \end{bmatrix}$$

$$\Rightarrow B_{eq} = AB + D(AB_C + B)$$

Quick solution: Therefore, if B in the line and B_C are both zero, B_{eq} must also be zero, so option 3 is the correct choice.

3.47. **Option 2 is correct**. Using equations (3.36), (3.39) and (3.40), we have:

$$100\,\text{MW} = \frac{3 \times \left(\dfrac{400\,\text{kV}}{\sqrt{3}} \right)^2}{Z_C} \Rightarrow Z_C = 1600 = \sqrt{\frac{L}{C}}, \quad \sqrt{LC} = \frac{10^{-8}}{3}$$

$$\Rightarrow \sqrt{C} \times 1600\sqrt{C} = \frac{10^{-8}}{3} \Rightarrow C = \frac{10^{-10}}{48}$$

3.48. **Option 3 is correct**.

Because the end of the circuit is open and the source voltage and load must be equal, the compensating impedance of the Z_p must remove the shunt capacitor. Therefore, the capacitor's admittance must be negative. Therefore, we have:

$$Y_P = -j0.001\,(\mho) \Rightarrow Z_P = j1000\,(\Omega)$$

$$Z_b = \frac{V_b^2}{S_b} = \frac{(200\,\text{kV})^2}{100\,\text{MVA}} = 400\,(\Omega) \Rightarrow Z_P = \frac{j1000}{400} = j2.5\,(\text{p.u.})$$

3.49. **Option 2 is correct.** Because this load is to the power of a SIL, it is a wave or natural load, and using equation (3.43) $V_X = V_R$ means the magnitude of the voltage is constant along the length of the line. There is only one question point: *effective* and *maximum* keywords.

In circuit lessons, we have $V_{max} = \sqrt{2}\, V_e$. So option 2 is correct.

3.50. **Option 3 is correct.** We have:

In short line: $\hat{V}_{S1} = \hat{V}_R + Z\hat{I}_R$

In the π model line we have:

$$\hat{V}_{S2} = \left(1 + \frac{ZY}{2}\right)\hat{V}_R + Z\hat{I}_R = \hat{V}_{S1} + \frac{ZY}{2}\hat{V}_R$$

$$\Rightarrow \left|\Delta\hat{V}_S\right| = \left|\hat{V}_{S2} - \hat{V}_{S1}\right| = \frac{|Z\|Y|}{2}\left|\hat{V}_R\right| = \frac{B\sqrt{R^2 + X^2}}{2}\left|V_R\right|$$

3.18 Two-Choice Questions (Yes – No) – 107 Questions

1. The capacitor effect is ignored in the short line.
2. The short line is <50 miles long.
3. Voltage on the short transmission line exceeds 69 kV.
4. The short line model has equal currents at the beginning and the end.
5. The transmission line can be viewed as bipolar.
6. The short transmission line in the hybrid matrix is $C = 0$ and $B = Z$.
7. Voltage regulation can be defined as a percentage change in voltage from no-load to full load at the beginning of a line.
8. Voltage regulation of the line can be defined in the form of

$$V_R\% = \frac{\hat{V}_{NL} - \hat{V}_{FL}}{\hat{V}_{FL}}.$$

9. $\begin{bmatrix} A & B \\ C & D \end{bmatrix}$ is the transmission matrix, so $\left|\hat{V}_{R(NL)}\right| = \frac{\left|\hat{V}_S\right|}{A}$.

10. In the short line $V_{R(NL)} = V_S$.
11. Voltage regulation is a criterion of the voltage drop and depends on the power factor of the load.
12. Voltage regulation will be greater than one in loads with a lagging power factor.

13. Voltage regulation will be negative in loads with a leading power factor.

14. A phasor diagram is shown below to illustrate the leading power factor of the short transmission line (Figure 3.12).

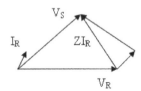

FIGURE 3.12
Question system 3.14.

15. The efficiency of a transmission line can be expressed as $\eta = \dfrac{P_{R(3\,ph)}}{P_{S(3\,ph)}}$.

16. Medium lines are lines with a length of more than 80 km and <150 miles.

17. In the T model of the medium line, half of the shunt capacitance is considered compacted on both sides.

18. The C in the transmission matrix of the medium line in the π model is $C = Y\left(1 + \dfrac{ZY}{2}\right)$.

19. The relation $AD - BC = 1$ is correct in the transmission line characteristic matrix.

20. The bipolar transmission lines are symmetrical.

21. Lines longer than 100 miles are called long lines.

22. It is essential to consider the effect of extensive parameters on a long-line model.

23. The propagation constant is $\gamma = \sqrt{zy}$.

24. The real part of the propagation constant is called the phase constant and its imaginary component is called the attenuation constant.

25. The characteristic impedance is $Z_C = \sqrt{\dfrac{z}{y}}$.

26. The voltage at any point of a long line is equal to the difference of the incident and reflected voltage waves.

27. The reflected voltage wave decreases as you move away from the end of the long line.

28. The imaginary part of the propagation constant becomes zero by ignoring line losses.

29. If the conductance and resistance of the line are zero, the magnitude of the propagation constant becomes equal to $\omega\sqrt{LC}$.

30. If the line is lossless, the characteristic impedance becomes pure inductance.

31. When $Z_{\text{Load}} = Z_C$, the loading is known as the natural load or SIL.

32. Power transmission velocity is equal to light velocity when the internal bonding flux of a conductor is ignored.

33. When the line is lossless, the A in the long line transmission matrix becomes $\cosh\left(\omega\sqrt{LC}x\right)$.

34. No-load long lines have a higher voltage on the send side than on the receive side.

35. The A in the transmission matrix decreases as the length of the lossless long line increases.

36. The surge impedance loading is the load in wave impedance at the nominal voltage.

37. In lossless surge impedance loading, the magnitude of the voltage and current at each point of the line is constant.

38. In lossless surge impedance loading, the reactive power of the line is zero.

39. In lossless surge impedance loading, inductance absorbs the same reactive power as capacitance generates.

40. SIL is an appropriate criterion for transmission line capacitance.

41. The SIL determines how much reactive power is required for a load.

42. Shunt reactors should be used for loads that are significantly greater than SIL.

43. Usually, the full load of a transmission line is different from its SIL.

44. Working load angles are normally between 35° and 45° to ensure sufficient stability margin.

45. The available transmission capability of a line is limited to its thermal loading limit and its stability limit.

46. The load angle (δ) decreases for a specified load when transformer and generator reactance are added to the line.

47. In a lossless line, the transmission power is

$$P_{3ph} = \frac{\left|\hat{V}_S(\text{p.u.})\right|\left|\hat{V}_R(\text{p.u.})\right|}{\sin\beta l}\sin\delta.$$

48. The thermal limit determines the maximum transmission power for short and medium lines.

49. Long lines are lines with a length of more than 200 km

50. The voltage curve along a transmission line with surge impedance loading is essentially flat.

51. On long transmission lines, loads much lower than SIL cause the voltage at the receiving end to decrease.

52. In long lines with equal voltage on both sides, the maximum voltage occurs in the middle.

53. In long lines with equal voltage on both sides, the maximum voltage occurs in the middle in the form of $V_m = \dfrac{|V_R|}{\cosh \dfrac{\beta l}{2}}$.

54. The percentage of voltage regulation can be obtained from the percentage of voltage drop.

55. You can reduce the voltage in the middle of the line by installing reactors on both sides.

56. Ratio $\dfrac{X_{Cser}}{X'}$ is commonly referred to as percentage compensation.

57. Percentage compensation ranges from 25% to 70%.

58. The transmission line admittance is the inverse of the line imped ance $z = \dfrac{1}{y}$.

59. In symmetric and balanced three-phase transmission lines, $|A| = |D| \le 1$.

60. The line voltage regulation can be calculated from the relation $R\% = \dfrac{P_R \cdot R + Q_R \cdot X}{V_R}$.

61. When two lines are connected in cascade, the equivalent transmission matrix is the sum of the transmission matrix.

62. By halving the transmission line, the (a) of the line half transmission matrix becomes equal to $a = \dfrac{\sqrt{1 + A}}{2}$.

63. The voltage regulation formula is: $R\% = \dfrac{V_{Rnl} - V_{Rfl}}{V_{Rfl}} \times 100$.

64. Voltage regulation for the lagging power factor is the minimum and for the leading power factor, it is the maximum.

65. A and D of the transmission lines matrix are dimensionless.

66. The dimensions of B and C are Ohm and Mho, respectively.

67. The unit of the attenuation value is Nepers per meter (Np/m).

68. The unit of the phase constant is radians per length meter (rad/m).
69. In the long line, the sending voltage is the same as the reflected voltage wave, and the receiving voltage is the same as the incident voltage wave.
70. When a line reaches its characteristic impedance, its reflected voltage and current waves become zero.
71. When a line reaches its characteristic impedance, it is called a flat or limited line.
72. The characteristic impedance in lines with a bundle conductor is higher.
73. The incident and reflected voltages on the no-load line are equal.
74. If high accuracy is not desired, a nominal π model can be used to display the long line.
75. In long lines, the line capacitance increases A in the transmission matrix.
76. In general, transient phenomena in transmission systems are caused by sudden changes in work conditions or system configuration.
77. In terms of dimensions, wave impedance is similar to pure resistance.
78. V and I are the same phases in a load of impulse impedance along a lossless line.
79. The voltage and current profile are non-flat in a load with impulse impedance.
80. A load of impulse impedance is an appropriate reference for evaluating and expressing the capability of that line.
81. High voltage transmission lines longer than 500 km must be terminated at an impedance close to their characteristic impedance.
82. The characteristic impedance should be reduced by adding compensators or the transmission voltage increased to increase the power transmission capability.
83. $\theta = \beta l$ is called the electrical length or angle of the line.
84. The largest voltage will be closer to the side that has the lower voltage on the long line, if the voltage on both sides is not equal.
85. The maximum power in the lagging power factor is greater.
86. In the case where the voltage magnitude on both sides of the line is equal, the voltage angle of the middle point of the line is equal to the middle of the angle between the voltages on each side.
87. Loadability of the line is defined as the line loading limit with thermal, voltage drop, and stability limits.
88. Loadability of the line is reduced with bundling.

89. As lagging current passes through the line inductance, the load side voltage increases.

90. As the leading current passes through the line inductance, the load side voltage increases.

91. Reactive power is transferred from one side whose voltage magnitude is lower to the other side in a two-machine system.

92. Active power losses increase with increased reactive power transmission.

93. In the T model, the C is $Z(1+ZY/2)$.

94. If the voltage regulation is negative, maybe the efficiency is negative.

95. The percentage of the voltage drop is defined only for no-load mode.

96. With a lagging load, voltage regulation becomes positive.

97. If the voltage regulation is negative, then the load is leading.

98. If the conductor is solid, $L \times C$ becomes a constant.

99. You can determine the characteristic impedance by knowing the values of A and B.

100. In long lines, static stability is more relevant than thermal stability.

101. In long lines, dynamic stability is more important than static stability.

102. The series and shunt reactances of the line become equal if the length of the line is a quarter of the wavelength.

103. When the line length is a quarter of the wavelength, a resonance occurs.

104. When the line length exceeds half the wavelength, the series impedance becomes capacitive.

105. The line shunt impedance becomes inductive if the line length is less than half the wavelength.

106. When we reduce the length of the long line model, we get the short line model.

107. When a shunt reactor is placed in the middle of the line, the voltage at the beginning, end, and middle of the line can be equalized.

3.19 Key Answers to Two-Choice Questions

1. Yes
2. Yes
3. No
4. Yes

5. Yes
6. No
7. No
8. No
9. No
10. Yes
11. Yes
12. Yes
13. Yes
14. Yes
15. Yes
16. Yes
17. No
18. No
19. Yes
20. Yes
21. No
22. Yes
23. Yes
24. No
25. Yes
26. No
27. No
28. No
29. Yes
30. No
31. Yes
32. Yes
33. No
34. No
35. Yes
36. Yes
37. Yes
38. Yes
39. Yes
40. Yes

41. No
42. No
43. Yes
44. Yes
45. Yes
46. No
47. Yes
48. Yes
49. No
50. Yes
51. No
52. Yes
53. No
54. No
55. Yes
56. Yes
57. Yes
58. No
59. Yes
60. No
61. No
62. No
63. Yes
64. No
65. Yes
66. Yes
67. Yes
68. Yes
69. No
70. Yes
71. No
72. No
73. Yes
74. Yes
75. No
76. Yes

77. Yes
78. Yes
79. No
80. Yes
81. Yes
82. Yes
83. Yes
84. No
85. No
86. Yes
87. Yes
88. No
89. No
90. No
91. No
92. Yes
93. No
94. No
95. No
96. Yes
97. Yes
98. No
99. Yes
100. Yes
101. Yes
102. Yes
103. Yes
104. Yes
105. No
106. Yes
107. Yes

4

Power Flow Analysis

Part One: Lesson Summary

4.1 Introduction

The most effective tool for power system operators and designers is to analyze and study power flow or load flow (LF). One of the oldest power system problems was solving LF quickly and accurately. There is still a lot of research being done to improve its speed and accuracy. This book chapter summarizes the relationships required for power system LF by Gauss-Seidel, Newton-Raphson, and DC LF methods. A total of 55 non-calculator problems are provided in this chapter, along with 59 true-false questions to help students better understand the concepts. This chapter focuses on:

Three major types of nodes or buses, Solution of nonlinear equations (Gauss, Newton), DC LF, Programming in the form of sparse matrices, Two-bus system LF (analytical method), Static voltage stability, LF based on voltage magnitude, Transformer with tap changers, Bus impedance matrix, Losses in simple power systems, LF of distribution systems.

The reader is required to read this lesson summary before answering the questions in this chapter, which emphasize key points.

4.2 Algebraic Equations

Target: It is commonly required to know the load and generation power for a LF. The voltage magnitudes and angles of all buses are output LF.

4.2.1 Required Relationships of Electrical Circuits

If $\hat{V}_i \triangleq V_i \angle \delta \triangleq e_i + jf_i$ is the voltage of the ith bus, $\hat{I}_{Gi}$ is the generator current of ith bus, $\hat{I}_{Di}$ is the load current of ith bus, and $\hat{I}_i$ is the injection current from the ith bus to the network, then we have:

The power generation of the *i*th bus is : $S_{Gi} = P_{Gi} + jQ_{Gi} = \hat{V}_i \cdot \hat{I}_{Gi}^*$ (4.1)

The power demand of the *i*th bus is : $S_{Di} = P_{Di} + jQ_{Di} = \hat{V}_i \cdot \hat{I}_{Di}^*$ (4.2)

The power and current injection from the *i*th bus to the network:

$$S_i = S_{Gi} - S_{Di} = P_{Gi} - P_{Di} + j(Q_{Gi} - Q_{Di}) = P_i + jQ_i = \hat{V}_i \cdot \hat{I}_i^*$$ (4.3)

$$\hat{I}_i = \hat{I}_{Gi} - \hat{I}_{Di}$$ (4.4)

4.2.2 The Line Model in the LF

In LF, the π model is usually used (Figure 4.1).
  Y_{Sij} = the line series admittance between bus i and bus j
  Y_{Pij} = the line shunt admittance between bus i and bus j

4.2.3 The Network Admittance Matrix

The relation between the vector of bus voltages and injected bus currents:

$$[I_{bus}] = [Y_{bus}][V_{bus}]$$ (4.5)

That $[V_{bus}]$ is the column vector of $n \times 1$ from the bus voltages and $[I_{bus}]$ is the column vector of $n \times 1$ from the injection bus current. The total number of buses is assumed to be n.
  The bus admittance matrix of a *n*-bus network:

$$[Y_{bus}] = \begin{bmatrix} Y_{11} & \cdots & Y_{1n} \\ \vdots & & \vdots \\ Y_{n1} & \cdots & Y_{nn} \end{bmatrix}_{n \times n}$$ (4.6)

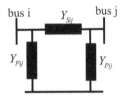

FIGURE 4.1
The π model of transmission line.

The diagonal and off-diagonal elements of the bus admittance matrix are:

$$Y_{ij} = -Y_{Sij} \quad i \neq j, \quad Y_{ii} = \sum_{j=1 \neq i}^{n} (Y_{Sij} + Y_{Pij}) \tag{4.7}$$

$$Y_{ij} = G_{ij} + jB_{ij} = |Y_{ij}| \angle \gamma_{ij}.$$

The relation between bus voltages and injected bus currents is:

$$\hat{I}_i = \sum_{j=1}^{n} (Y_{ij} \hat{V}_j) \tag{4.8}$$

4.2.4 The Power Equations (Polar Form)

Combining relations (4.3), (4.7), and (4.8) and eliminating the current, we obtain the following power equation:

$$P_i = \sum_{j=1}^{n} V_i V_j |Y_{ij}| \cos(\delta_i - \delta_j - \gamma_{ij}) \triangleq f_{Pi}(V, \delta) \tag{4.9}$$

$$\Rightarrow P_i = V_i^2 G_{ii} + \sum_{j=1 \neq i}^{n} V_i V_j |Y_{ij}| \cos(\delta_i - \delta_j - \gamma_{ij})$$

$$\Rightarrow P_i = V_i^2 G_{ii} + V_i \sum_{j=1 \neq i}^{n} V_j \left(G_{ij} \cos(\delta_i - \delta_j) + B_{ij} \sin(\delta_i - \delta_j) \right)$$

$$Q_i = \sum_{j=1}^{n} V_i V_j |Y_{ij}| \sin(\delta_i - \delta_j - \gamma_{ij}) \triangleq f_{Qi}(V, \delta) \tag{4.10}$$

$$\Rightarrow Q_i = -V_i^2 B_{ii} + \sum_{j=1 \neq i}^{n} V_i V_j |Y_{ij}| \sin(\delta_i - \delta_j - \gamma_{ij})$$

$$\Rightarrow Q_i = -V_i^2 B_{ii} + V_i \sum_{j=1 \neq i}^{n} V_j \left(G_{ij} \sin(\delta_i - \delta_j) - B_{ij} \cos(\delta_i - \delta_j) \right)$$

4.2.5 The Power Equations (Cartesian Form)

In Cartesian form, equations (4.9) and (4.10) are also related: $\hat{V}_i \triangleq e_i + jf_i$ and $Y_{ij} = G_{ij} + jB_{ij}$.

$$\text{We define}: \quad a_{ij} = G_{ij}e_j - B_{ij}f_j, \quad b_{ij} = G_{ij}f_j + B_{ij}e_j \tag{4.11}$$

$$P_i = e_i \sum_{j=1}^{n} a_{ij} + f_i \sum_{j=1}^{n} b_{ij} \triangleq f_{pi}\left(e_i, f_i\right) \tag{4.12}$$

$$Q_i = f_i \sum_{j=1}^{n} a_{ij} - e_i \sum_{j=1}^{n} b_{ij} \triangleq f_{qi}\left(e_i, f_i\right) \tag{4.13}$$

4.3 Three Major Types of Nodes or Buses

We define six variables for each bus: $P_{Gi}, Q_{Gi}, P_{Di}, Q_{Di}, V_i, \delta_i$
    Two variables in each bus are unknown, while the rest are known.

1. The **Load Bus or (PQ)**: In this bus, the voltage magnitude (V_i) and the voltage angle (δ_i) are unknown.
2. The **Voltage-Controlled Bus or Regulated Bus or (PV)**: In this bus, the generation reactive power (Q_{Gi}) and the voltage angle (δ_i) are unknown.
    The voltage magnitude of this bus is specified (V_{isp}).
3. The **Reference or Slack Bus (SL) or Swing Bus**: In this bus, the generation real power (P_{Gi}) and the generation reactive power (Q_{Gi}) are unknown. In this bus usually $V = 1$, $\delta = 0$.

4.3.1 The Bus Numbering

The first bus is usually the reference bus and buses 2 to m are voltage-controlled buses if the total number of buses is n.
    Load buses (PQ buses) are numbered from bus $m+1$ to bus n.
    It is assumed that this issue has no generality, for the sake of simplicity and problem-solving.

4.3.2 The Tips of Voltage Control Bus

Voltage-controlled buses (PV) are controlled by a reactive power controller or compensator. This controller defines Q_{max} as the maximum generation reactive power and Q_{min} as the minimum generation reactive power.
    The voltage magnitude in voltage-controlled buses (PV) is constant and known as ($V_i = V_{isp}$) if Q_i of generation is calculated from equation (4.10) within the range $\left(Q_{i\min} \leq Q_i \leq Q_{i\max}\right)$.

Otherwise, this bus is converted to a load bus (PQ) and the generation Q_i remains within its limits of $(Q_i = Q_{imin})$ and/or $(Q_i = Q_{imax})$ and the voltage magnitude of this bus becomes unknown.

4.4 Solution of Nonlinear Equations (Gauss, Newton) $F(x) = 0$

A. Gauss:

The nonlinear equation of $F(x) = 0$ is converted into an equation of $x = f(x)$ to obtain the following iterative relation:

$$x^{(r+1)} = f\left(x^{(r)}\right) \tag{4.14}$$

In the above iterative relation, $x^{(r)}$ represents the value of x in the rth iteration and $x^{(0)}$ represents the initial estimate.

B. Newton:

Using Taylor's series expansion:

$$F(x) = 0 \Rightarrow F\left(x^{(0)} + \Delta x^{(0)}\right) = 0$$

$$F(x^{(0)}) + \Delta x^{(0)} \left[\frac{\partial F}{\partial x}\right]^{(0)} + \frac{1}{2}\left[\Delta x^{(0)}\right]^2 \left[\frac{\partial^2 F}{\partial x^2}\right]^{(0)} + \cdots = 0$$

In the absence of $\left[\Delta x^{(0)}\right]^2$ and higher degrees of $\Delta x^{(0)}$, we have:

$$F(x^{(0)}) + \Delta x^{(0)} \left[\frac{\partial F}{\partial x}\right]^{(0)} \simeq 0 \Rightarrow \Delta x^{(0)} = -\left(\left[\frac{\partial F}{\partial x}\right]^{(0)}\right)^{-1} \times F\left(x^{(0)}\right)$$

The iterative formula for the Newton method : $x^{(r+1)} = x^{(r)} + \Delta x^{(r)}$ (4.15)

4.4.1 The Advantages and Disadvantages of the Methods (Gauss-Newton)

The advantages of the Gauss method
The relations are very simple.
The disadvantages of the Gauss method
It may take a long time to reach an answer and it may diverge.
The advantages of the Newton method
It takes a short time to reach the answer. The method is stable and converges most of the time. The number of iterations is independent of the number of buses.

The disadvantages of the Newton method
These relationships are difficult because of the use of derivatives.

4.4.2 Acceleration Coefficient

$$x^{(r+1)} = x^{(r)} + \alpha \times \Delta x^{(r)} \tag{4.16}$$

In general $\alpha = 1.5$ is assumed. After a few iterations of using the acceleration coefficient, α will equal one. This will ensure that the equation will not pass away from the original answer and the iterative equation will not oscillate.

4.4.3 The Gauss-Jacobi or Gauss-Jacobian Iterative Method

Using equations (4.3) and (4.8) and the Gauss method we have:

$$S_i = \hat{V}_i \cdot \hat{I}_i^* = \hat{V}_i \left(\sum_{j=1}^{n} Y_{ij} \cdot \hat{V}_j \right)^* \Rightarrow \frac{S_i^*}{\hat{V}_i^*} = Y_{ii}\,\hat{V}_i + \sum_{j=1,\neq i}^{n} Y_{ij} \cdot \hat{V}_j$$

$$\Rightarrow \hat{V}_i = \frac{P_i - jQ_i}{Y_{ii} \cdot \hat{V}_i^*} - \sum_{j=1,\neq i}^{n} \frac{Y_{ij}}{Y_{ii}} \cdot \hat{V}_j$$

By writing the above relation as an iterative form, we get the following Gauss-Jacobi iterative equation:

$$\hat{V}_i^{(r+1)} = \frac{P_i - jQ_i}{Y_{ii} \cdot \hat{V}_i^{*(r)}} - \sum_{j=1,\neq i}^{n} \frac{Y_{ij}}{Y_{ii}} \hat{V}_j^{(r)}, \quad i = 2,\ldots,n \tag{4.17}$$

AS $\hat{V}_i^{(r)}$ is the ith bus voltage in the rth iteration and $\hat{V}_i^{(0)}$ is the initial estimate of the voltage, usually assumed to be $1\angle 0$ and, as the first bus is usually the reference bus, the equation (4.17) for $i = 2$ onwards is used.

4.4.4 The Gauss-Seidel Iterative Method

Gauss-Seidel can be obtained by converting Σ (equation 4.17) into two parts as follows. Compared to the Gauss-Jacobi method, this method responds faster. The Seidel method is superior to Jacobi because $\hat{V}_i^{(r+1)}$ is closer to the answer than $\hat{V}_i^{(r)}$ if the problem has a solution.

$$\hat{V}_i^{(r+1)} = \frac{P_i - jQ_i}{Y_{ii} \cdot \hat{V}_i^{*(r)}} - \sum_{j=1}^{i-1} \frac{Y_{ij}}{Y_{ii}} \hat{V}_j^{(r+1)} - \sum_{j=i+1}^{n} \frac{Y_{ij}}{Y_{ii}} \hat{V}_j^{(r)}, \quad i = 2,\ldots,n \tag{4.18}$$

4.4.5 The Newton-Raphson Iterative Method

The following definitions are required for the Newton-Raphson algorithm:

$$
\text{The main equation: } F(x) = 0 = \begin{bmatrix} P - f_P^{(r)} \\ Q - f_Q^{(r)} \end{bmatrix}_{(2n-2)\times 1} \tag{4.19}
$$

$$
\text{Active and reactive power vector: } Q = \begin{bmatrix} Q_2 \\ \vdots \\ Q_n \end{bmatrix}_{(n-1)\times 1}, \quad P = \begin{bmatrix} P_2 \\ \vdots \\ P_n \end{bmatrix}_{(n-1)\times 1}
$$

$$\tag{4.20}$$

$$
\text{The Jacobian matrix: } [J]^{(r)} \triangleq \left[\frac{\partial F}{\partial x}\right]^{(r)}_{(2n-2)\times(2n-2)} \tag{4.21}
$$

$$
\text{The iterative formula: } X^{(r+1)} = X^{(r)} + \Delta X^{(r)}, \quad [J]^{(r)} \cdot \Delta X^{(r)} = \Delta U^{(r)} \tag{4.22}
$$

The definition of known values is:

$$
\Delta U^{(r)} = \begin{bmatrix} \Delta P_2^{(r)} & \cdots & \Delta P_n^{(r)} & \Delta Q_2^{(r)} & \cdots & \Delta Q_n^{(r)} \end{bmatrix}^T = \begin{bmatrix} \Delta P^{(r)} & \Delta Q^{(r)} \end{bmatrix}^T
$$

$$\tag{4.23}$$

The definition of unknown values is:

$$
\Delta X^{(r)} = \begin{bmatrix} \Delta \delta_2^{(r)} & \cdots & \Delta \delta_n^{(r)} & \Delta V_2^{(r)} & \cdots & \Delta V_n^{(r)} \end{bmatrix}^T = \begin{bmatrix} \Delta \delta^{(r)} & \Delta V^{(r)} \end{bmatrix}^T
$$

$$\tag{4.24}$$

That the superscript of T means transpose.
   The definition of $\Delta P_i^{(r)}$ and $\Delta Q_i^{(r)}$:

$$
\Delta P_i^{(r)} = P_i - f_{Pi}\left(\delta^{(r)}, V^{(r)}\right) \text{ and } \Delta Q_i^{(r)} = Q_i - f_{Qi}\left(\delta^{(r)}, V^{(r)}\right) \tag{4.25}
$$

It is possible to calculate $f_{Pi}\left(\delta^{(r)}, V^{(r)}\right)$ and $f_{Qi}\left(\delta^{(r)}, V^{(r)}\right)$ using equations (4.9) and (4.10), and P_i and Q_i are known using equation (4.3).

The Jacobian matrix components are:

$$[J]^{(r)} = \begin{bmatrix} J_1 & J_2 \\ J_3 & J_4 \end{bmatrix}^{(r)} = \begin{bmatrix} \dfrac{\partial f_P}{\partial \delta} & \dfrac{\partial f_P}{\partial V} \\ \dfrac{\partial f_Q}{\partial \delta} & \dfrac{\partial f_Q}{\partial V} \end{bmatrix}^{(r)}_{(2n-2)\times(2n-2)} \qquad (4.26)$$

The extensive iterative method of the Jacobian matrix:

$$\begin{bmatrix} J_1 & J_2 \\ J_3 & J_4 \end{bmatrix}^{(r)} \begin{bmatrix} \Delta\delta \\ \Delta V \end{bmatrix}^{(r)} = \begin{bmatrix} \Delta P \\ \Delta Q \end{bmatrix}^{(r)} \qquad (4.27)$$

4.4.6 Jacobian Matrix Calculations

The diagonal elements of the Jacobian matrix $(i = 2,\ldots,n)$:

$$J_{1ii} = \frac{\partial f_{Pi}}{\partial \delta_i} = -V_i \sum_{k=1\neq i}^{n} V_k \cdot |Y_{ik}| \sin(\delta_i - \delta_k - \gamma_{ik})$$

$$= -f_{Qi} - V_i^2 B_{ii} \simeq -Q_i - V_i^2 B_{ii}$$

$$J_{2ii} = \frac{\partial f_{Pi}}{\partial V_i} = 2V_i |Y_{ii}| \cos(\gamma_{ii}) + \sum_{k=1\neq i}^{n} V_k |Y_{ik}| \cos(\delta_i - \delta_k - \gamma_{ik})$$

$$= \frac{f_{Pi}}{V_i} + V_i G_{ii} \simeq \frac{P_i}{V_i} + V_i G_{ii} \qquad (4.28)$$

$$J_{3ii} = \frac{\partial f_{Qi}}{\partial \delta_i} = V_i \sum_{k=1\neq i}^{n} V_k \cdot |Y_{ik}| \cos(\delta_i - \delta_k - \gamma_{ik}) = f_{Pi} - V_i^2 G_{ii} \simeq P_i - V_i^2 G_{ii}$$

$$J_{4ii} = \frac{\partial f_{Qi}}{\partial V_i} = -2V_i |Y_{ii}| \sin(\gamma_{ii}) + \sum_{k=1\neq i}^{n} V_k |Y_{ik}| \sin(\delta_i - \delta_k - \gamma_{ik})$$

$$= \frac{f_{Qi}}{V_i} - V_i B_{ii} \simeq \frac{Q_i}{V_i} - V_i B_{ii}$$

There is no row or column corresponding to the slack bus in the four matrixes for J_1, J_2, J_3, J_4.

You can calculate f_{Pi} and f_{Qi} using equations (4.9) and (4.10) and P_i and Q_i are known using equation (4.3).

The off-diagonal elements of the Jacobian matrix as Polar and Cartesian forms are $i \neq k$:

$$J_{1ik} = \frac{\partial f_{Pi}}{\partial \delta_k} = V_i \cdot V_k \cdot |Y_{ik}| \sin(\delta_i - \delta_k - \gamma_{ik}) = a_{ik} f_i - b_{ik} e_i$$

$$J_{2ik} = \frac{\partial f_{Pi}}{\partial V_k} = V_i \cdot |Y_{ik}| \cos(\delta_i - \delta_k - \gamma_{ik}) = \frac{a_{ik} e_i + b_{ik} f_i}{V_k} = \frac{a_{ik} e_i + b_{ik} f_i}{\sqrt{e_k^2 + f_k^2}}$$

$$(4.29)$$

$$J_{3ik} = \frac{\partial f_{Qi}}{\partial \delta_k} = -V_i \cdot V_k \cdot |Y_{ik}| \cos(\delta_i - \delta_k - \gamma_{ik}) = -a_{ik} e_i - b_{ik} f_i$$

$$J_{4ik} = \frac{\partial f_{Qi}}{\partial V_k} = V_i \cdot |Y_{ik}| \sin(\delta_i - \delta_k - \gamma_{ik}) = \frac{a_{ik} f_i - b_{ik} e_i}{V_k} = \frac{a_{ik} f_i - b_{ik} e_i}{\sqrt{e_k^2 + f_k^2}}.$$

4.4.7 Use of Symmetry of Jacobian Matrix (the Modified Jacobian Matrix)

So that the components of the Jacobian matrix (equations 4.28 and 4.29) have many similarities. Calculating the Jacobian matrix can be sped up by changing the equation of equation (4.27) to the following equation. This modified Jacobian matrix is called the modified Jacobian matrix.

$$\begin{bmatrix} J_1 & J_2' \\ J_3 & J_4' \end{bmatrix}^{(r)} \begin{bmatrix} \Delta\delta \\ \dfrac{\Delta V}{V} \end{bmatrix}^{(r)} = \begin{bmatrix} \Delta P \\ \Delta Q \end{bmatrix}^{(r)} \qquad (4.30)$$

Now, we can simplify equations (4.28) and (4.29) as follows.
The diagonal elements of the modified Jacobian matrix ($i=2, \ldots n$) are:

$$J_{1ii} = -f_{Qi} - V_i^2 B_{ii} \simeq -Q_i - V_i^2 B_{ii}, \quad J_{4ii}' = f_{Qi} - V_i^2 B_{ii} \simeq Q_i - V_i^2 B_{ii}$$

$$J_{2ii}' = f_{Pi} + V_i^2 G_{ii} \simeq P_i + V_i^2 G_{ii}, \qquad J_{3ii} = f_{Pi} - V_i^2 G_{ii} \simeq P_i - V_i^2 G_{ii}$$

$$(4.31)$$

The off-diagonal elements of the modified Jacobian matrix as Polar and Cartesian form are $i \neq k$:

$$J_{1ik} = J_{4ik}' = V_i \cdot V_k \cdot |Y_{ik}| \sin(\delta_i - \delta_k - \gamma_{ik}) = a_{ik} f_i - b_{ik} e_i$$

$$J_{2ik}' = -J_{3ik} = V_i \cdot V_k \cdot |Y_{ik}| \cos(\delta_i - \delta_k - \gamma_{ik}) = a_{ik} e_i + b_{ik} f_i$$

$$(4.32)$$

When using the modified Jacobian matrix, be careful that the unknowns (equation 4.24) and the iterative equation of equation (4.22) must be changed as follows.

The new unknown variables of the modified Jacobian matrix are $\Delta X'$:

$$\Delta X'^{(r)} = \begin{bmatrix} \Delta\delta_2^{(r)} & \cdots & \Delta\delta_n^{(r)} & \dfrac{\Delta V_2^{(r)}}{V_2^{(r)}} & \cdots & \dfrac{\Delta V_n^{(r)}}{V_n^{(r)}} \end{bmatrix}^T = \begin{bmatrix} \Delta\delta^{(r)} & \dfrac{\Delta V^{(r)}}{V^{(r)}} \end{bmatrix}^T$$

$$(4.33)$$

The iterative formula for the modified Jacobian matrix is:

$$X'^{(r+1)} = X'^{(r)} + \Delta X'^{(r)}, \quad [J']^{(r)} \cdot \Delta X'^{(r)} = \Delta U^{(r)} \tag{4.34}$$

4.4.8 The Newton-Raphson Method Algorithm

A. The initial estimate is $r = 0$ (usually $\delta_i^{(0)} = 0$ and $V_i^{(0)} = 1$) $X^{(r)} = \begin{bmatrix} \delta^{(r)} \\ V^{(r)} \end{bmatrix}$

B. Calculate $\Delta P_i^{(r)}$ and $\Delta Q_i^{(r)}$ using equation (4.25) and then calculate $\Delta U^{(r)}$ using equation (4.23).

C. Calculate $[J]^{(r)}$ using equations (4.26), (4.28) and (4.29) or calculate $[J']^{(r)}$ using equations (4.31) and (4.32).

D. Calculate $\Delta X^{(r)}$ using equation (4.24) or calculate the $\Delta X'^{(r)}$ using equation (4.33).

E. Calculate $X^{(r+1)}$ using equation (4.22) or calculate the $X'^{(r+1)}$ using equation (4.34).

F. Repeat the above equations $(r \rightarrow r + 1)$ until the norm of $\Delta U^{(r)}$ or $\Delta X^{(r)}$ is less than the maximum error value (ε).

4.4.9 Definition of Norm or $\| \ \|$

$$\text{If we have: } X = \begin{bmatrix} X_1 & \cdots & X_n \end{bmatrix}_{1 \times n}$$

$$m\text{th-norm: } \|X\|_m = \sqrt[m]{\sum_{i=1}^{n} |X_i|^m} \tag{4.35}$$

$$\text{Euclidean norm: } \|X\|_2 = \sqrt[2]{\sum_{i=1}^{n} |X_i|^2} \tag{4.36}$$

$$\text{Infinity norm: } \|X\|_\infty = \max_i |X_i|$$

4.4.10 Decoupled Load Flow (DLF)

The matrices of J_2 (or J_2') and J_3 in J (or J') become ignored. The speed of this method is ten times faster.

$$\begin{cases} [\Delta P] \simeq [J_1][\Delta \delta] \\ [\Delta Q] \simeq [J_4][\Delta V] \end{cases} \quad \text{and} \quad \text{or} \quad \begin{cases} [\Delta P] \simeq [J_1][\Delta \delta] \\ [\Delta Q] \simeq [J_4']\left[\dfrac{\Delta V}{V}\right] \end{cases} \tag{4.37}$$

Note that the matrix rank of J_1, is $n-1$ and the matrix rank of J_4 or J_4' is $n-1-m$.

4.4.11 Fast Decoupled Load Flow

To simplify the equations, in addition to removing the matrices of J_2 (or J_2') and J_3, the matrices of J_1 and J_4 are also assumed.

$$\delta_i \simeq \delta_j, \quad \gamma_{ik} = 90°, \quad \gamma_{ii} = -90°, \quad Q_i \ll B_{ii}V_i^2 \tag{4.38}$$

$$\sin(\delta_i - \delta_k - \gamma_{ik}) \simeq -1, \quad \sin(\gamma_{ii}) \simeq -1, \quad \cos(\delta_i - \delta_k - \gamma_{ik}) \simeq 0, \quad \cos(\gamma_{ii}) \simeq 0$$

The matrixes J_1 and J_4' are simplified as follows:

$$J_{1ii} = J_{4ii}' = -V_i^2 B_{ii} \quad \text{and} \quad J_{1ik} = J_{4ik}' = -V_i V_k B_{ik} = -e_i e_k B_{ik} \tag{4.39}$$

If we divide ΔP and ΔQ by V in equation (4.37), and in the residual matrices, we set V_i equal to 1, we arrive at the following very simple relationship:

$$\left[\frac{\Delta P}{V}\right] = -[B'][\Delta \delta] \tag{4.40a}$$

$$\left[\frac{\Delta Q}{V}\right] = -[B''][\Delta V] \tag{4.40b}$$

The speed of this method is very high, but accuracy is low. In equation (4.40), the $[B']$ and $[B'']$ matrix is the imaginary part of the admittance matrix, except that the row and column related to the slack bus are removed. In addition, the matrix rank of $[B'']$ at (equation 4.40b), is $n-m-1$ (the row and column related to the voltage control buses have also been removed from that).

4.5 The Direct LF or DC Load Flow

With the following simplifications in the famous relationship $P_{ij} = V_i V_j \sin(\delta_i - \delta_j)/x_{ij}$, this method can achieve the LF commonly used in contingency ranking.

$$V_i = 1, \; R_{ij} = 0, \; \sin(\delta_i - \delta_j) \approx (\delta_i - \delta_j) \tag{4.41}$$

By using the following relation:

$$P_i = \sum_{j=1}^{n} P_{ij} = \sum_{j=1}^{n} \frac{\delta_i - \delta_j}{x_{ij}} \tag{4.42}$$

The DC LF is summarized as follows:

$$[P] = [B][\delta] \tag{4.43}$$

Note that the row and column related to the Slack bus have been removed and

$$B_{ii} = \sum_{j=1 \neq i}^{n} \frac{1}{x_{ij}}, \; B_{ij} = -\frac{1}{x_{ij}} \tag{4.44}$$

We can also obtain equation (4.43) by assuming $V_i = 1 \Rightarrow \Delta V = 0$ from the relation of equation (4.40) (FDLF).

$$[B] = -[B'] \tag{4.45}$$

To solve the equation of equation (4.43), we have:

$$[X] = [B]^{-1} \Rightarrow [\Delta\delta] = [X][\Delta P] \tag{4.46}$$

When the power in the ith bus increases the ΔP_i, the angle of the other buses (except for the reference, which is constant) is obtained from the following relation:

$$\Delta\delta_j = X_{ji} \cdot \Delta P_i \tag{4.47}$$

The X_{ij} obtained from the matrix of X is not equal to the x_{ij} of the line reactance between the bus of i and j.

4.6 Programming in the Form of SPARSE Matrices

A large number of buses in the power system will result in large admittance and Jacobian matrices. In each row, they probably won't have more than five or six nonzero arrays, and these arrays will be scattered around the matrix's main diagonal. To reduce the amount of memory required, only nonzero arrays are stored in memory. When Z is the number of zeros in the matrix and n is the dimension of the matrix, sparsity is defined as follows:

$$\text{Sparsity: } S = \frac{Z}{n^2} \tag{4.48}$$

$$\text{The nonzero elements: } Z = (1-s)n^2 \tag{4.49}$$

Here is a simple example of sparsity.
  If A is a sample matrix:

$$A = \begin{bmatrix} 5 & -9 & -10 & 0 \\ -9 & 6 & 0 & 0 \\ -10 & 0 & 7 & -11 \\ 0 & 0 & -11 & 8 \end{bmatrix}$$

The vector of diagonal elements: $Z_{\text{diag}} = \begin{bmatrix} 5 & 6 & 7 & 8 \end{bmatrix}$
  The vector of off-diagonal elements:

$$Z_{\text{offd}} = \begin{bmatrix} -9 & -10 & -9 & -10 & -11 & -11 \end{bmatrix}$$

The row of off-diagonal elements: $I_{\text{row}} = \begin{bmatrix} 1 & 1 & 2 & 3 & 3 & 4 \end{bmatrix}$

The column of off-diagonal elements: $I_{\text{col}} = \begin{bmatrix} 2 & 3 & 1 & 1 & 4 & 3 \end{bmatrix}$

We can apply matrix symmetry as follows if we want to use it:

$I_R = \begin{bmatrix} 1 & 1 & 3 \end{bmatrix}$, $I_c = \begin{bmatrix} 2 & 3 & 4 \end{bmatrix}$, $Z_{\text{offd}} = \begin{bmatrix} -9 & -10 & -11 \end{bmatrix}$,

$Z_{\text{diag}} = \begin{bmatrix} 5 & 6 & 7 & 8 \end{bmatrix}$

4.7 Two-Bus System Load Flow (Analytical Method)

In the Figure 4.2, the LF of a simple two-bus system can be calculated accurately without the use of iterative methods.

$$V_2^2 = \frac{\left(V_1^2 - 2Qx\right) \pm \sqrt{V_1^2\left(V_1^2 - 4Qx\right) - 4x^2 P^2}}{2} \tag{4.50}$$

A positive sign gives us a stable answer, and a negative sign gives us an unstable answer and we have:

$$\delta = \delta_1 - \delta_2 = \sin^{-1}\left(\frac{P \times x}{V_1 \times V_2}\right) \tag{4.51}$$

If the line model in Figure 4.2 is more accurate and the admittance matrix of the system is known in the form of: $\left[Y_{bus}\right]_{2 \times 2}$, V_2^2 is obtained as follows:

$$V_2^2 = \frac{-\beta \pm \sqrt{\beta^2 - 4\alpha\gamma}}{2\alpha} \tag{4.52}$$

Where $\alpha = \left|Y_{22}\right|^2$, $\gamma = \left(S_2\right)^2$, $\beta = 2\left(G_{22}P_2 - B_{22}Q_2\right) - \left|Y_{12}\right|^2$ $\quad$ (4.53)

4.7.1 The LF of the Two-Bus System (Gauss-Seidel Method)

There is no difference between Gauss-Jacobi and Gauss-Seidel in the two-bus system, whereas in the simple system of Figure 4.2, the following summary transforms:

$$\hat{V}_2^{(r+1)} = \frac{-x\left(Q + jP\right)}{\hat{V}_2^{*(r)}} + 1 \tag{4.54}$$

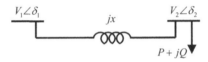

FIGURE 4.2
A simple two-bus system.

4.8 Static Voltage Stability

Briefly, the inability of the system to maintain voltage is referred to as voltage instability. The LF cannot determine the bus voltage in this case. This may be caused by high load power (especially reactive power) or increased resistivity when parallel lines are lost.

The voltage stability condition for the simple system of Figure 4.2 is specified as follows (the radicand must be positive):

$$V_1^2 \left(V_1^2 - 4Qx \right) - 4x^2 P^2 \geq 0 \tag{4.55}$$

The radicand is zeroed when the stability boundary is reached, and two values of V_2 become equal:

$$V_2 = \sqrt{\frac{V_1^2 - 2Qx}{2}} \tag{4.56}$$

For a more accurate line model in Figure 4.2, the voltage stability condition is as follows:

$$\beta^2 - 4\alpha\gamma \geq 0 \text{ and } \beta \leq 0 \left(\alpha, \beta, \text{ and } \gamma \text{ using equation} (4.53) \right) \tag{4.57}$$

4.9 LF Based on Voltage Magnitude

If we ignore the voltage angle in the power system, we get the following linear relation:

$$Q_i \cos\left(\delta_i\right) - P_i \sin\left(\delta_i\right) + A_{ii} + \underbrace{\sum A_{ij} V_j}_{\substack{\text{slack \&} \\ \text{generator} \\ \text{buses}}} + \underbrace{\sum A_{ij}}_{\substack{\text{other} \\ \text{load} \\ \text{buses}}}$$

$$= \Delta V_i \left(Q_i \cos\left(\delta_i\right) - P_i \sin\left(\delta_i\right) - A_{ii} \right) - \underbrace{\sum A_{ij} \Delta V_j}_{\substack{\text{other} \\ \text{load} \\ \text{buses}}} \tag{4.58}$$

That: $A_{ij} = B_{ij} \cos\left(\delta_j\right) + G_{ij} \sin\left(\delta_j\right)$ \tag{4.59}

4.10 Transformer with Tap Changers

In the above system, the transformer has a tap of α and a series impedance of y_t. Transformers and taps are defined as follows (transformer is per-unit) (Figure 4.3).

$$\hat{I}_i = -\alpha^* \hat{I}_j \text{ and } \hat{V}_x = \frac{\hat{V}_j}{\alpha} \tag{4.60}$$

The transformer admittance matrix can be calculated as follows:

$$\begin{bmatrix} \hat{I}_i \\ \hat{I}_j \end{bmatrix} = \begin{bmatrix} y_t & -\dfrac{y_t}{\alpha} \\ -\dfrac{y_t}{\alpha^*} & \dfrac{y_t}{\alpha^2} \end{bmatrix} \begin{bmatrix} \hat{V}_i \\ \hat{V}_j \end{bmatrix} \tag{4.61}$$

When α is real, we can represent the transformer with the following asymmetric π model (Figure 4.4).

$$\text{The } \pi \text{ model of the transformer with tap similar to Figure 4.4} \tag{4.62}$$

Note, if $\alpha > 1$, the shunt admittance on the bus of j (tap side) is capacitance and on the bus of i (non-tap side) is inductance, and if $\alpha < 1$, the shunt admittance on the bus of j (tap side) is inductance and the bus of i (non-tap side) is capacitance.

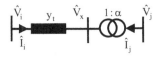

FIGURE 4.3
Transformer with tap changers.

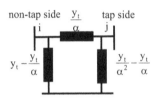

FIGURE 4.4
Transformer model in the form of π model.

4.11 Bus Impedance Matrix

The impedance matrix obtained from the inverse of the admittance matrix is frequently used ($Z_{bus} = Y_{bus}^{-1}$).

Added a short circuit impedance Z_f to the kth bus results in:

$$\hat{I}_f = \frac{\hat{V}_k^o}{Z_f + Z_{kk}}, \quad \hat{V}_i^f = \hat{V}_i^o - Z_{ik} \times \hat{I}_f, \quad \hat{I}_{ij} = \frac{\hat{V}_i^f - \hat{V}_j^f}{z_{ij}} \qquad (4.63)$$

The voltage changes on the ith bus:

$$\Delta V_i = \left| \hat{V}_i^f - \hat{V}_i^o \right| = \frac{|Z_{ik}| \times V_k^o}{|Z_f + Z_{kk}|} \qquad (4.64)$$

A short circuit current is $\hat{I}_f$, the voltage of buses after a short circuit is $\hat{V}_i^f$, the current of lines after the short circuit is $\hat{I}_{ij}^f$, $\hat{V}_i^o$ the voltage of the buses before fault, z_{ij} is the line series impedance between bus i and j and Z_{ik} and Z_{kk} are determined from the impedance matrix.

When impedances with the following conditions are added to the network, the impedance matrix will change as follows:

A. If the bus of $n+1$ with an impedance of Z_b is connected to the kth bus.

$$\left[Z_{bus}^{new} \right] = \begin{bmatrix} Z_{bus} & Z_{ok} \\ Z_{ko} & Z_b + Z_{kk} \end{bmatrix} \qquad (4.65)$$

where Z_{ko} is the kth row and Z_{ok} is the kth column of the $[Z_{bus}]$ matrix.

B. If the kth bus is connected to the earth with impedance of Z_f:

In matrix form: $\left[Z_{bus}^{new} \right] = [Z_{bus}] - \dfrac{Z_{ok} \times Z_{ko}}{Z_{kk} + Z_f} \qquad (4.66)$

In array form: $Z_{ij}^{new} = Z_{ij} - \dfrac{Z_{ik} \times Z_{kj}}{Z_{kk} + Z_f} \qquad (4.67)$

In this case, equations (4.63) and (4.64) are also true.

C. If the impedance of Z_f is connected between bus h and bus k:

In matrix form:

$$\left[Z_{bus}^{new} \right] = \left[Z_{bus} \right] - \frac{(Z_{ok} - Z_{oh})(Z_{ko} - Z_{ho})}{Z_{hh} + Z_{kk} + Z_f - 2Z_{kh}} \tag{4.68}$$

In array form:

$$Z_{ij}^{new} = Z_{ij} - \frac{(Z_{ik} - Z_{ih})(Z_{kj} - Z_{hj})}{Z_{hh} + Z_{kk} + Z_f - 2Z_{kh}} \tag{4.69}$$

The current passing through the impedance of Z_f:

$$\hat{I}_f = \frac{\hat{V}_k^o - \hat{V}_h^o}{Z_{hh} + Z_{kk} + Z_f - 2Z_{kh}} \tag{4.70}$$

The voltage of the bus after addition of Z_f:

$$\hat{V}_i^f = \hat{V}_i^o - (Z_{ik} - Z_{ih}) \times \hat{I}_f \tag{4.71}$$

The voltage change in the ith bus:

$$\Rightarrow \Delta V_i = \left| \hat{V}_i^f - \hat{V}_i^o \right| = \left| \frac{(Z_{ik} - Z_{ih})(\hat{V}_k^o - \hat{V}_h^o)}{Z_{hh} + Z_{kk} + Z_f - 2Z_{kh}} \right| \tag{4.72}$$

The voltage changes ratio in the ith bus:

$$\frac{\Delta V_i}{\Delta V_j} = \left| \frac{Z_{ik} - Z_{ih}}{Z_{jk} - Z_{jh}} \right| \tag{4.73}$$

4.11.1 Bus Admittance Matrix

4.11.1.1 The Node Elimination

Load buses can be removed so that no elements are attached. Taking I_X as a subvector of the vector current of these buses, we get:

$$[I] = [Y_{bus}][V] \Rightarrow \begin{bmatrix} I_A \\ I_X \end{bmatrix} = \begin{bmatrix} K & L \\ L^T & M \end{bmatrix} \begin{bmatrix} V_A \\ V_X \end{bmatrix} \tag{4.74}$$

$$[I_X] = 0 \Rightarrow [I_A] = [Y_{red}][V_A], \quad [Y_{red}] = K - LM^{-1}L^T \tag{4.75}$$

4.12 Losses in Simple Power Systems

One generator and one line.
  Two generators and one line.
  For Figures 4.5 and 4.6, we have:

$$P_L = B_{11}P_1^2 \tag{4.76}$$

That:

$$B_{11} = \frac{R}{\left(V_1 \cos\varphi_1\right)^2}$$

Two generators and two lines (Figure 4.7):

$$P_L = B_{11}P_1^2 + B_{22}P_2^2 \tag{4.77}$$

That: $B_{11} = \dfrac{R_1}{\left(V_1 \cos\varphi_1\right)^2}$ and $B_{22} = \dfrac{R_2}{\left(V_2 \cos\varphi_2\right)^2}$

Two generators and three lines (Figure 4.8):

FIGURE 4.5
One generator and one line.

FIGURE 4.6
Two generators and one line.

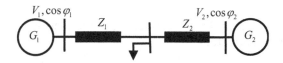

FIGURE 4.7
Two generators and two lines.

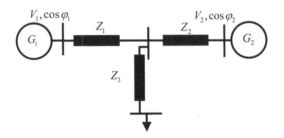

FIGURE 4.8
Two generators and three lines.

$$P_L = B_{11}P_1^2 + B_{22}P_2^2 + 2B_{12}P_1P_2 \tag{4.78}$$

$$\text{That:}\quad B_{11} = \frac{R_1}{\left(V_1\cos\varphi_1\right)^2} + \frac{R_3}{\left(V_3\cos\varphi_3\right)^2}$$

$$B_{12} = \frac{R_3}{\left(V_3\cos\varphi_3\right)^2} \quad\text{and}\quad B_{22} = \frac{R_2}{\left(V_2\cos\varphi_2\right)^2} + \frac{R_3}{\left(V_3\cos\varphi_3\right)^2}$$

4.13 LF of Distribution Systems

Definition:
$\Delta U\% \triangleq$ The percentage of voltage drop
$\Delta U \triangleq$ The voltage drop
$A \triangleq$ The cross-sectional area in mm^2
$\Delta P\% \triangleq$ The percentage of power loss or losses
$I \triangleq$ The current in amperes
$l \triangleq$ The line length in meters
$\rho \triangleq$ The electrical resistivity or special electrical resistivity in Ohm-meter
$\sigma \triangleq \dfrac{1}{\rho}$ The conductivity or special conductivity in Mho/m, for copper is 56
and for aluminum is 35 in $10^6\,\Omega$m

$$\text{The electrical resistance:}\quad R = \frac{l}{\sigma A} \tag{4.79}$$

- The direct current (DC) is connected to the load via the same send and return conductors:

$$\text{The percentage of voltage drop:}\quad \Delta U\% = \frac{200lI}{\sigma AU} = \frac{200lP}{\sigma AU^2} \tag{4.80}$$

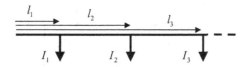

FIGURE 4.9
The distribution of load along a line.

$$\text{The power loss drop:} \quad \Delta P = \frac{2lI^2}{\sigma A} = \frac{2lP^2}{\sigma AU^2} \tag{4.81}$$

- This is the distribution of load along a line, as shown in Figure 4.9. The percentage of the voltage drop:

$$\Delta U\% = \frac{200\sum_{i=1}^{n} l_i I_i}{\sigma AU} = \frac{200\sum_{i=1}^{n} l_i P_i}{\sigma AU^2} \tag{4.82}$$

- The single-phase AC line is connected to the load via the same send and return conductors

$$\text{The percentage of the voltage drop:} \quad \Delta U\% = \frac{200lI}{\sigma AU}\cos(\varphi) = \frac{200lP}{\sigma AU^2} \tag{4.83}$$

$$\text{The power loss drop:} \quad \Delta P = \frac{2lP^2}{\sigma A(U\cos(\varphi))^2} \tag{4.84}$$

$$\text{The percentage of the power loss drop:} \quad \Delta P\% = \frac{200lP}{\sigma A(U\cos(\varphi))^2} \tag{4.85}$$

- The multi-branched three-phase (Radial transmission line) (Figure 4.10)

$$\text{The voltage drop:} \quad \Delta U = \frac{1}{U}\sum_{i=1}^{n}(r_i P_i + x_i Q_i) \tag{4.86}$$

$$\text{The percentage of the voltage drop:} \quad \Delta U\% = \frac{100}{U^2}\sum_{i=1}^{n}(r_i P_i + x_i Q_i) \tag{4.87}$$

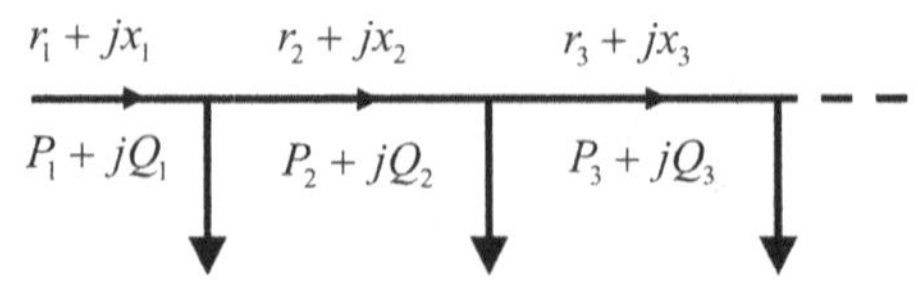

FIGURE 4.10
The radial transmission line.

- AC line connecting both ends of the power supply (with the same voltage) (Figure 4.11)

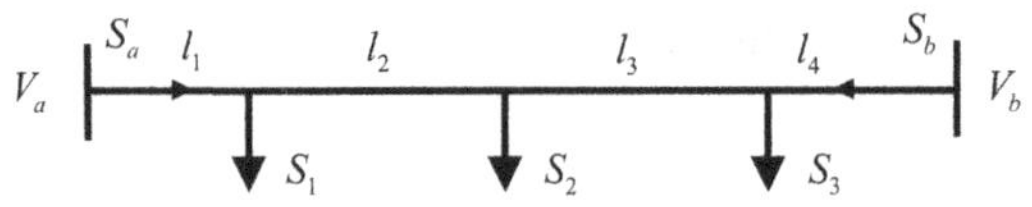

FIGURE 4.11
AC line connecting both ends of the power supply (Equation 4.88).

$$\text{The injection power:}\quad S_a = \frac{S_3 l_4 + S_2 (l_3 + l_4) + S_1 (l_2 + l_3 + l_4)}{l_1 + l_2 + l_3 + l_4}$$

$$S_b = \sum S_i - S_a \tag{4.88}$$

The injection current:

$$\hat{I}_a = \frac{\hat{I}_3 l_4 + \hat{I}_2 (l_3 + l_4) + \hat{I}_1 (l_2 + l_3 + l_4)}{l_1 + l_2 + l_3 + l_4}, \quad \hat{I}_b = \sum \hat{I}_i - \hat{I}_a \tag{4.89}$$

4.13.1 The Definition of Deep Point

It is the point in the network with the greatest voltage drop and the current direction that is changed.

The voltage drop at the deep point (if the load 2 is the deep point)

$$\Delta U = \frac{l_1 \hat{I}_a + l_2 \left(\hat{I}_a - \hat{I}_1 \right)}{\sigma A} \tag{4.90}$$

This is the power drop if load 2 is the deep point.

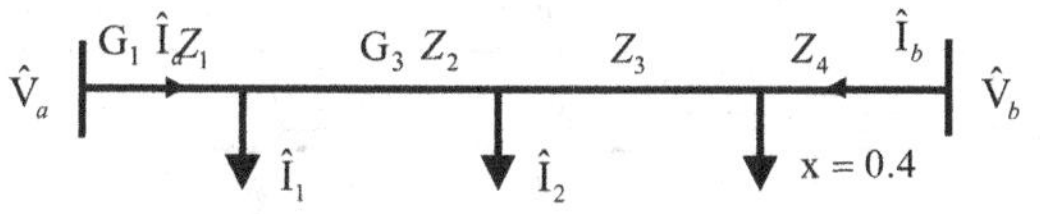

FIGURE 4.12
AC line connecting both ends of the power supply (Equation 4.93).

$$\Delta P = \frac{l_1 S_a^2 + l_2 \left(S_a - S_1\right)^2 + l_3 \left(S_b - S_3\right)^2 + l_4 S_b^2}{\sigma A U^2} \qquad (4.91)$$

The percentage of power loss drop: $\quad \Delta P\% = \dfrac{100\Delta P}{\displaystyle\sum_i P_i + \Delta P} \qquad (4.92)$

- For the following system, we have (Figure 4.12):
  The injection current:

$$\hat{I}_a = \frac{\hat{V}_a - \hat{V}_b + \hat{I}_3 Z_4 + \hat{I}_2 \left(Z_3 + Z_4\right) + \hat{I}_1 \left(Z_2 + Z_3 + Z_4\right)}{Z_1 + Z_2 + Z_3 + Z_4},$$

$$\hat{I}_b = \sum \hat{I}_i - \hat{I}_a \qquad (4.93)$$

Voltage drop at the deep point from the side of V_a (if the load 2 is the deep point).

$$\Delta U = Z_1 \hat{I}_a + Z_2 \left(\hat{I}_a - \hat{I}_1\right) \qquad (4.94)$$

All the above relationships assume that the lines have the same cross-sectional area. We can use the following relation if the areas are unequal.

$$l' = l \frac{A'}{A} \qquad (4.95)$$

Where A' is the largest cross-sectional area, l' is the updated virtual length of the line, and l and A are the actual length and cross-sectional area of the line.

- For the following system, we have (Figure 4.13):
  The injection power:

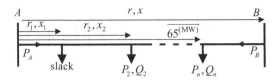

FIGURE 4.13
AC line connecting both ends of the power supply (Equation 4.96).

$$P_B = \frac{1}{r}(P_1 r_1 + P_2 r_2 + \cdots + P_n r_n) = \sum_{i=1}^{n} P_i - P_A$$

$$P_A = \frac{1}{r}(P_1(r - r_1) + P_2(r - r_2) + \cdots + P_n(r - r_n))$$

$$Q_B = \frac{1}{x}(Q_1 x_1 + Q_2 x_2 + \cdots + Q_n x_n) = \sum_{i=1}^{n} Q_i - Q_A \qquad (4.96)$$

$$Q_A = \frac{1}{x}(Q_1(x - x_1) + Q_2(x - x_2) + \cdots + Q_n(x - x_n))$$

4.13.2 The Continuous Load

In a radial line with resistance of r per unit length, and current or power of I and P, the voltage drop at a continuous load with a length of l equals:

$$\Delta V = 2rIl^2 = 2r\frac{P}{V}l^2 \qquad (4.97)$$

By simple integration, the relation (4.97) can be proven.

Part Two: Answer Question

4.14 Four-Choice Questions – 55 Questions

4.1. Using the DC LF, if the power of generator 1 increases by 10%, how many radians does the voltage angle of bus 1 (δ_1) increase (Figure 4.14)? $V_i = 1$, $S_{base} = 100$ MVA

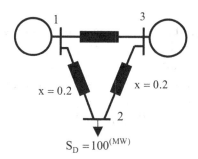

FIGURE 4.14
Question system 4.1.

   1. 0.013   2. 0.065   3. 0.026   4. 0.0065

4.2. How many radians does the voltage angle of bus 1 increase if the power of generator 1 increases by 10% and the power of load bus 2

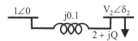

FIGURE 4.15
Question system 4.3.

   increases by 6.5%?
   1. 0   2. 0.013   3. 0.0195   4. 0.0065

4.3. What is the maximum value of Q in the following system until the LF has an answer (Figure 4.15)?
   1. 2   2. 2.1   3. 1.9   4. 1.8

4.4. In question 4.3, if $Q=2$, what is the stable answer to the voltage magnitude?
   1. $\dfrac{\sqrt{2}}{2}$ 2. $\sqrt{0.4}$ 3. 0.8   4. 0.9

4.5. In question 4.3, if $Q=2$, what is the unstable answer to the voltage magnitude?

1. $\sqrt{0.1}$ 2. $\sqrt{0.2}$ 3. $\sqrt{0.3}$ 4. $\sqrt{0.4}$

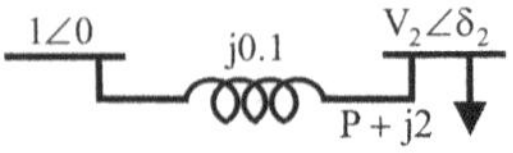

$$1\angle 0 \quad j0.1 \quad V_2\angle\delta_2 \quad P+j2$$

FIGURE 4.16
Question system 4.6.

4.6. What is the magnitude of voltage at the border of voltage stability in the following simple system (Figure 4.16)?
    1. $\sqrt{0.6}$ 2. $\sqrt{0.5}$ 3. $\sqrt{0.4}$ 4. $\sqrt{0.3}$

4.7. What is the magnitude of voltage on load bus 2, at first repetition, if we solve the LF of the following system using the Gauss-Seidel method and $\hat{V}_2^{(0)} = 1\angle 0$ (Figure 4.17)?

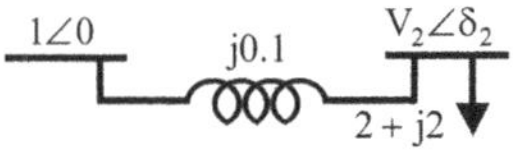

$$1\angle 0 \quad j0.1 \quad V_2\angle\delta_2 \quad 2+j2$$

FIGURE 4.17
Question system 4.7.

1. $\sqrt{0.86}$ 2. $\sqrt{0.8}$ 3. $\sqrt{0.68}$ 4. $\sqrt{0.6}$

4.8. In the following system, there are two transformers with a tap changer. What is the magnitude of y_{44} in the admittance matrix (Figure 4.18)?
    1. 15.25   2. 14   3. 13   4. 15

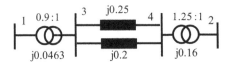

FIGURE 4.18
Question system 4.8.

4.9. What is the $\Delta\delta_2$ in radians when bus 3 is the voltage control in FDLF (Figure 4.19)?

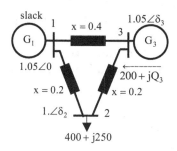

FIGURE 4.19
Question system 4.9.

$$\Delta P_2 = 0.1, \ S_b = 100 \text{ MVA}, \ \Delta P_3 = 0.105$$

1. 0.025   2. 0.0255   3. 0.035   4. 0.0355

4.10. If $\Delta Q_2 = 1.01$ and ΔQ_3 is unknown in the previous question system, what is ΔV_2? (Assume $V_2 = 1.01$, $V_3 = 1.04$ and the other parameters remain the same.)

1. 0.1   2. −0.1   3. 0.101   4. −0.101

4.11. If we solve the Problem 4.9 using the DC LF method, what is $\Delta \delta_2$ in radians?

1. 0.0355   2. 0.035   3. 0.025   4. 0.0255

4.12. What is the Thevenin impedance from bus 1 if the admittance matrix is in front form? $Y_{\text{bus}} = j \begin{bmatrix} -3 & 1 & 2 \\ 1 & -4 & 3 \\ 2 & 3 & -6 \end{bmatrix} (\text{p.u.})$

1. $\dfrac{15}{11}$ 2. $\dfrac{11}{15}$ 3. 7   4. $\dfrac{11}{6}$

4.13. For question 4.12, if $\hat{V}_1 = 1.1\angle 0$ and $\hat{V}_2 = 1.2\angle 0$ and an inductance with admittance of $\dfrac{11}{9}$ is connected to bus 1, the voltage of bus 2 changes to V_{22}, then what is V_{22}?

1. 0.55   2. 0.5   3. 0.65   4. 0.6

4.14. How much Thevenin reactance can be seen from bus 2 in the system of question 4.12 if bus 3 is removed and the system is converted to two buses?

1. $\dfrac{15}{11}$ 2. $\dfrac{11}{15}$ 3. 7   4. $\dfrac{14}{11}$

4.15. How much magnitude of y_{11} in the updated admittance matrix in question 4.12 if bus 3 is removed and the system is converted to two buses?

1. $\dfrac{5}{2}$ 2. $\dfrac{7}{3}$ 3. $\dfrac{2}{3}$ 4. $\dfrac{14}{11}$

4.16. In a three-bus system, we have $\hat{V}_2 = 1.2\angle 0$, $\hat{V}_1 = \hat{V}_3 = 1.1\angle 0$ and the impedance matrix is in following form.

What is the magnitude of the current passing through a capacitor connected between bus 1 and 2 with a reactance of 0.15?

$$Z_{bus} = j \begin{bmatrix} 0.2 & 0.15 & 0.1 \\ 0.15 & 0.3 & 0.15 \\ 0.1 & 0.15 & 0.25 \end{bmatrix} (\text{p.u.})$$

   1. 0.3   2. 0.15   3. 1.5   4. 2

4.17. In question 4.16, what is the magnitude of the bus voltage of 3?

   1. 0.9   2. 0.95   3. 1   4. 1.05

4.18. In question 4.16, what is the voltage change ratio of bus 2 to bus 3?

   1. 1   2. 2   3. 3   4. 4

4.19. Consider a three-wire AC three-phase system and a two-wire DC system with equal resistance. How much loss occurs in the AC to DC system at equal transmission power and voltage peak?

   1. $\dfrac{3}{4}$ 2. $\dfrac{3}{2}$ 3. $\dfrac{4}{3}$ 4. $\dfrac{2}{3}$

4.20. In the case of a three-wire AC three-phase and a two-wire DC system with the same transmission power, losses, and resistance, what is the ratio of insulation level of the DC system to the AC system?

   1. $\dfrac{3}{2}$ 2. $\dfrac{\sqrt{3}}{2}$ 3. $\dfrac{2}{3}$ 4. $\dfrac{2}{\sqrt{3}}$

4.21. When an AC system is connected to a DC system, and both systems are per-unit, what is the ratio of DC base current to AC base current?

   1. 1   2. $\sqrt{2}$ 3. 2   4. $\sqrt{3}$

4.22. A three-phase system loses 81 MW, a DC system's insulation level is 0.9, and transmission power and line resistance are the same for both systems. What is the loss of the DC system in MW?

   1. 81   2. 90   3. 75   4. 72

4.23. What is the current I_{dc} in per unit for the front system (Figure 4.20)?

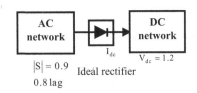

FIGURE 4.20
Question system 4.23.

   1. 0.6   2. 0.72   3. 0.8   4. 0.86

4.24. Suppose generators 1 and 2 generate 100 MW of reactive power in the following system. The losses are then 5 MW. If $P_1 = 80$ MW and $P_2 = 120$ MW, what are the losses in MW (Figure 4.21)?

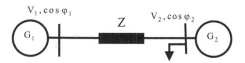

FIGURE 4.21
Question system 4.24.

   1. 2.5   2. 3.2   3. 5   4. 6.25

4.25. The voltage of all buses in the following system is 1 p.u. and $P_1 = \dfrac{5}{6}$ and $P_2 = 0.8$. When the power of the first generator changes by 2%, what percentage do the losses change (Figure 4.22)?

   1. 2   2. 1.8   3. 3.6   4. 7.2

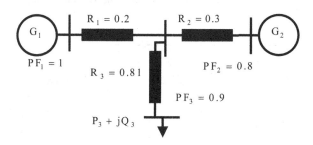

FIGURE 4.22
Question system 4.25.

4.26. A ten-bus system has 40 non-parallel lines. What is the sparsity of the admittance matrix?

   1. 0.1   2. 0.2   3. 0.3   4. 0.4

4.27. A ten-bus symmetric system has 40 non-parallel lines. How much memory is used to store the admittance matrix in the normal state compared to the sparse state?

   1. $\dfrac{55}{130}$ 2. $\dfrac{130}{55}$ 3. $\dfrac{45}{100}$ 4. $\dfrac{100}{45}$

4.28. In a symmetric ten-bus system, what is the maximum number of lines for which the Sparsity method of storing the admittance matrix is affordable?

   1. 5   2. 15   3. 30   4. 40

4.29. What is the percentage of voltage drop in the following system (Figure 4.23) if the cross-sectional area of the main cable (two aluminum strings) is $A = \dfrac{1000\,\text{mm}^2}{7}$?

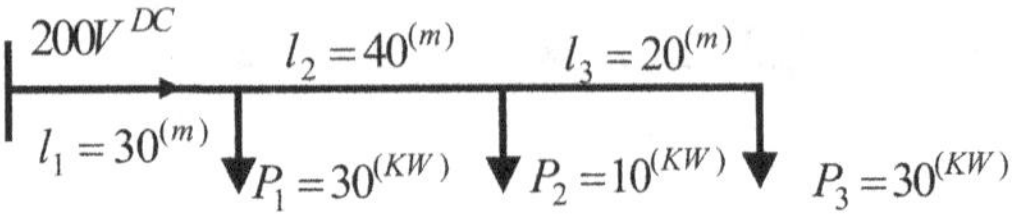

$$200V^{DC}$$
$$l_1 = 30^{(m)} \qquad l_2 = 40^{(m)} \qquad l_3 = 20^{(m)}$$
$$P_1 = 30^{(KW)} \qquad P_2 = 10^{(KW)} \qquad P_3 = 30^{(KW)}$$

FIGURE 4.23
Question system 4.29.

    1. 4.3   2. 1.7   3. 6.8   4. 2.42

4.30. For question 4.29, if the cross section is not the same, calculate the percentage of voltage drop again. $A_1 = 150\,\text{mm}^2$, $A_2 = 50\,\text{mm}^2$, $A_3 = 25\,\text{mm}^2$

    1. 9   2. 10   3. 11   4. 12

4.31. What is the voltage drop of the following system (Figure 4.24)?

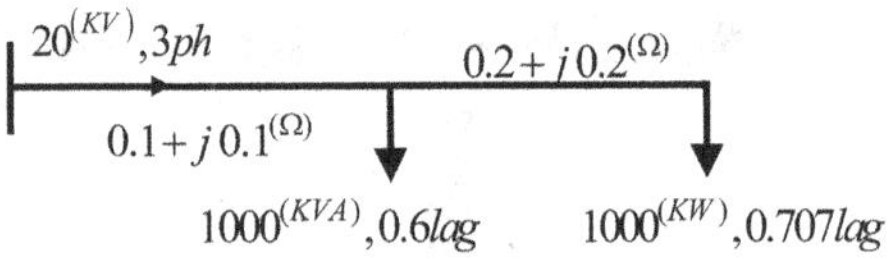

$$20^{(KV)},3ph$$
$$0.2 + j\,0.2^{(\Omega)}$$
$$0.1 + j\,0.1^{(\Omega)}$$
$$1000^{(KVA)},0.6lag \qquad 1000^{(KW)},0.707lag$$

FIGURE 4.24
Question system 4.31.

    1. 9   2. 18   3. 37   4. 72

4.32. In the following Figure 4.25 system, what is the current I_A?

$$I_A$$
$$l \qquad l \qquad l \qquad l$$
$$50A \qquad 50A \qquad 50A$$

FIGURE 4.25
Question system 4.32.

    1. 50   2. 60   3. 75   4. 80

4.33. In question 4.32, instead of three loads of 50 A, we have n loads with (I) amperes of current at equal intervals of (l), what is the current of I_A?

    1. $\dfrac{I(n+1)}{2}$ 2. $\dfrac{In(n+1)}{2}$ 3. nI 4. $\dfrac{nI}{2}$

4.34. If n loads of power P are distributed uniformly along a radial line with resistance r per unit length and equal intervals of l, what is the voltage drop in this system with voltage U?

1. $\dfrac{rlPn(n+1)}{2U}$ 2. $\dfrac{n^2 rlP}{U}$ 3. $\dfrac{n(n+1)rlP}{U}$ 4. $\dfrac{(n+1)^2 rlP}{U}$

4.35. According to question 4.34, if the line connecting both ends of the power supply, and n is even, what is the voltage drop?

1. $\dfrac{rlPn(n+1)}{8U}$ 2. $\dfrac{rlPn(n+2)}{8U}$ 3. $\dfrac{rlP(n+1)}{4U}$ 4. $\dfrac{rlPn(n+2)}{4U}$

4.36. According to question 4.34, if the line connects both ends of the power supply, and n is odd, what is the voltage drop?

1. $\dfrac{rlP(n+1)^2}{4U}$ 2. $\dfrac{rlPn^2}{4U}$ 3. $\dfrac{rlP(n+1)^2}{8U}$ 4. $\dfrac{rlPn(n+1)}{8U}$

4.37. The line in Figure 4.26 connects both ends of the power supply. If the bus voltage is 200 V, the voltage drop percentage is 5%, and the special electrical resistivity is 50, what is the cross-sectional area in square millimeters (Figure 4.26)?

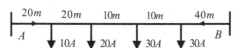

FIGURE 4.26
Question system 4.37.

1. 2 2. 4.6 3. 7.4 4. 3.7

4.38. What is the magnitude of the voltage drop in volts in the following ring network (Figure 4.27)? (The impedances are in ohms and the currents are in amps.)

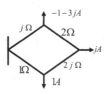

FIGURE 4.27
Question system 4.38.

1. $\dfrac{\sqrt{5}}{3}$ 2. $\dfrac{2\sqrt{5}}{3}$ 3. $\dfrac{4}{3}$ 4. $\dfrac{2}{3}$

4.39. If $\hat{V}_i \triangleq e_i + jf_i$ and $Y_{ij} = jB_{ij}$, what is $\dfrac{\partial P_i}{\partial e_i}$ in the LF in the Cartesian form?

1. $-e_i \displaystyle\sum_{j=1}^{n} B_{ij} f_j$ 2. $f_i \displaystyle\sum_{j=1}^{n} B_{ij} e_j$ 3. $-\displaystyle\sum_{j=1 \ne i}^{n} B_{ij} e_j$ 4. $-\displaystyle\sum_{j=1 \ne i}^{n} B_{ij} f_j$

4.40. What is the voltage difference in bus 3 from the Gauss-Seidel method with the Gaussian Jacobi method in the second iteration if the information of a three-bus system is as follows? $\left(\left| \Delta \hat{V}_3^{(2)} \right| \right)$:

$$Y_{bus} = j \begin{bmatrix} -3 & 1 & 2 \\ 1 & -4 & 3 \\ 2 & 3 & -6 \end{bmatrix}, \ \hat{V}_1 = 1\angle 0, \ \left\{ \begin{array}{l} \hat{V}_2^{(1)} = 1.1\angle 0 \\ \hat{V}_3^{(1)} = 1.15\angle 30° \end{array} \right. , \ \left| \begin{array}{l} \hat{V}_2^{(2)} = 1.2\angle 0 \\ \hat{V}_3^{(2)} = 1.2\angle 30° \end{array} \right.$$

1. 0.05   2. 0.1   3. 0.01   4. 0.02

4.41. In the Jacobian matrix, what is array $\dfrac{\partial f_{P_2}}{\partial \delta_2} P_{22}$ in the following Figure 4.28 system? Bus 3 has been voltage-controlled and $S_b = 100$ MVA

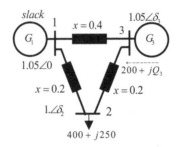

FIGURE 4.28
Question system 4.41.

1. 5.25   2. 10   3. 12.5   4. 10.5

4.42. In question 4.41, if bus 3 is not voltage-controlled and we have:
$\hat{V}_3 = 1\angle 60°$, $\hat{V}_2 = 1\angle 30°$, $\hat{V}_1 = 1\angle 0$, $Q_3 = 200$ MVAr

and the remaining parameters are constants. The Newton-Raphson LF is as follows:

$$\begin{bmatrix} J_1 & J_2' \\ J_3 & J_4' \end{bmatrix}^{(r)} \begin{bmatrix} \Delta\delta \\ \dfrac{\Delta V}{V} \end{bmatrix}^{(r)} = \begin{bmatrix} \Delta P \\ \Delta Q \end{bmatrix}^{(r)}$$

Then what is the value of this expression?

(The first row and column array of the matrix J_1) divided by
(The first row and the second column array of the matrix J_3)

1. 5   2. –5   3. 12.5   4. 2.5

4.43. According to Newton-Raphson LF, the error value is as follows:

$$\Delta X = \begin{bmatrix} \Delta \delta_2 & \Delta \delta_3 & \Delta V_2 & \Delta V_3 \end{bmatrix} = \begin{bmatrix} 0.15 & -0.2 & -0.1 & 0.05 \end{bmatrix}$$

In that case, what is the ratio of the error with norm 1 compared to

the error with an infinity norm? $\left(\dfrac{\|\Delta X\|_1}{\|\Delta X\|_\infty} = ? \right)$

1. 2.5   2. 0.5   3. –0.5   4. 0

4.44. What is the matrix determinant of Z_{bus} in the following network (Figure 4.29)?

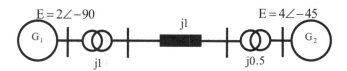

FIGURE 4.29
Question system 4.44.

1. –0.5   2. +0.5   3. –0.2   4. +0.2

4.45. The reactance of each transmission line L_1 and L_2 of the power system in Figure 4.30 is 0.5 p.u. What is the generation power of bus 2 (P_2) in terms of p.u.? Consider the DC LF with $\pi=3$.

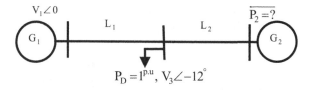

FIGURE 4.30
Question system 4.45.

1. 0.2   2. 0.25   3. 0.6   4. 0.75

4.46. This is the result of the LF and Z_{bus} matrix of a power system. If a capacitor with a reactance of $X_c = 3.4$ p.u. is connected to bus 4, what will be the voltage on bus 4?

Bus	1	2	3	4
V(p.u.)	1.02∠0°	0.98∠ – 15°	1.05∠10°	0.9∠0°

$$Z_{bus} = j \begin{bmatrix} 0.2 & 0.15 & 0.25 & 0.24 \\ 0.15 & 0.3 & 0.13 & 0.14 \\ 0.25 & 0.13 & 0.5 & 0.25 \\ 0.24 & 0.14 & 0.25 & 0.4 \end{bmatrix} (p.u.)$$

1. 0.95   2. 0.98   3. 1.02   4. 1.2

4.47. The fuel cost curves for the two power plants are as follows:

$$C_1 = 1000 + 50P_{G_1} + 0.0075P_{G1}^2$$

$$C_2 = 3000 + 45P_{G_2} + 0.005P_{G2}^2$$

What is the economic dispatch of 1000 MW of load between these two power plants?

1. 900 and 100 2. 750 and 250   3. 600 and 400   4. 200 and 800

4.48. In the following distribution network, what is the voltage drop at the "deep point" from bus A (Figure 4.31)? (The distance is equal (100 m), and each line has the same resistance 10^{-4}).

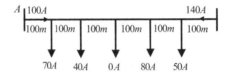

FIGURE 4.31
Question system 4.48.

1. 5.7   2. 1.3   3. 3.9   4. 2.3

4.49. This is the Y_{bus} admittance matrix for a two-bus system. In this case, buses (1) and (2) are connected by the same parallel lines. If one of the lines is disconnected from bus 1 and connected to earth, what is

called Y_{bus}? $Y_{bus} = \begin{bmatrix} -j10 & j10 \\ j10 & -j10 \end{bmatrix}$

1. $\begin{bmatrix} -j5 & j5 \\ j5 & -j10 \end{bmatrix}$ 2. $\begin{bmatrix} -j20 & j20 \\ j20 & -j20 \end{bmatrix}$

3. $\begin{bmatrix} -j20 & j5 \\ j5 & -j10 \end{bmatrix}$ 4. $\begin{bmatrix} -j5 & j5 \\ j5 & -j5 \end{bmatrix}$

4.50. In the following network, the single-phase line connecting the two ends of the power supply should have a resistance of 0.2 Ω. What is the minimum line voltage (deep point) (Figure 4.32)?

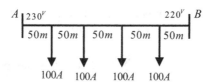

FIGURE 4.32
Question system 4.50.

1. 218   2. 212   3. 215   4. 225

4.51. In the following network, RT is the regulation transformer with a turn ratio of $a = \dfrac{\hat{V}_3}{\hat{V}_2} = |a| \angle \alpha$ and is ideal. What is the Y_{bus} for this network (Figure 4.33)?

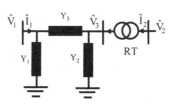

FIGURE 4.33
Question system 4.51.

1. $\begin{bmatrix} Y_1 + Y_3 & -\dfrac{Y_3}{a} \\ -\dfrac{Y_3}{a^*} & \dfrac{Y_2 + Y_3}{|a|^2} \end{bmatrix}$

2. $\begin{bmatrix} Y_1 + Y_3 & -a^* Y_3 \\ -a Y_3 & |a|^2 (Y_2 + Y_3) \end{bmatrix}$

3. $\begin{bmatrix} Y_1 + Y_3 & -a Y_3 \\ -a Y_3 & a^2 (Y_2 + Y_3) \end{bmatrix}$

4. $\begin{bmatrix} Y_1 + Y_3 & -a Y_3 \\ -a^* Y_3 & |a|^2 (Y_2 + Y_3) \end{bmatrix}$

4.52. Assume that the admittance matrix elements in the LF problem are in the form of $Y_{ij} = G_{ij} + jB_{ij}$ for a three-phase power network consisting of n buses. In bus Kth, we have voltage $\hat{V}_K = V_K \angle \delta_K$, injection active power P_k, and injection reactive power Q_k. Which of the following relationships is true for bus ith?

1. $\dfrac{Q_i \cos \delta_i - P_i \sin \delta_i}{V_i} = -\sum\limits_{j=1}^{n} V_j \left(B_{ij} \cos \delta_j + G_{ij} \sin \delta_j \right)$

2. $\dfrac{Q_i \cos \delta_i + P_i \sin \delta_i}{V_i} = -\sum\limits_{j=1}^{n} V_j \left(B_{ij} \cos \delta_j + G_{ij} \sin \delta_j \right)$

3. $\dfrac{Q_i \sin \delta_i - P_i \cos \delta_i}{V_i} = -\sum\limits_{j=1}^{n} V_j \left(B_{ij} \cos \delta_j + G_{ij} \sin \delta_j \right)$

4. $\dfrac{Q_i \sin \delta_i + P_i \cos \delta_i}{V_i} = -\sum\limits_{j=1}^{n} V_j \left(B_{ij} \cos \delta_j + G_{ij} \sin \delta_j \right)$

4.53. A three-bus power system has a bus voltage of 2 equal to $1.2\,\text{p.u.}\angle 0°$ and a matrix of Z_{bus} equal to the following. A 2.7 p.u. inductance is connected to bus 2, what is the magnitude change in bus voltage in

$$\text{bus 3? } Z_{bus} = j \begin{bmatrix} 0.2 & 0.15 & 0.1 \\ 0.15 & 0.3 & 0.15 \\ 0.1 & 0.15 & 0.25 \end{bmatrix} (\text{p.u.})$$

1. 0.075   2. 0.06   3. 0.12   4. 0.15

4.54. Following is a diagram of a three-phase power network with four buses, in which all resistances have been ignored and the reactance of each element is 0.1 p.u. The admittance matrix of this network in the equivalent LF problem is $[Y_{bus}]$. If lines 2–4 and 3–1 were disconnected from the network, the admittance matrix would look like $\left[Y_{bus\,new} \right] = \left[Y_{bus} \right] + \left[\Delta Y_{bus} \right]$.

Which is $[\Delta Y_{bus}] = \begin{bmatrix} A & B \\ C & D \end{bmatrix}$ (Figure 4.34)?

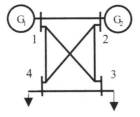

FIGURE 4.34
Question system 4.54.

1. $A = D = B = C = \begin{bmatrix} j10 & 0 \\ 0 & j10 \end{bmatrix}$

2. $A = D = -B = -C = \begin{bmatrix} j10 & 0 \\ 0 & j10 \end{bmatrix}$

3. $-A = -D = B = C = \begin{bmatrix} j10 & 0 \\ 0 & j10 \end{bmatrix}$

4. $A = D = B = C = \begin{bmatrix} -j10 & 0 \\ 0 & -j10 \end{bmatrix}$

4.55. A copper concentration factory operates on DC voltage, and its diagram is shown in the following. Which admittance matrix is used for analyzing the LF problem in this factory (Figure 4.35)?

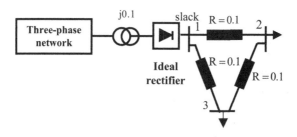

FIGURE 4.35
Question system 4.55.

1. $\begin{bmatrix} 20 & -10 & -10 \\ -10 & 20 & -10 \\ -10 & -10 & 20 \end{bmatrix}$

2. $\begin{bmatrix} 20 - j10 & -10 & -10 \\ -10 & 20 & -10 \\ -10 & -10 & 20 \end{bmatrix}$

3. $\begin{bmatrix} -20 + j10 & 10 & 10 \\ 10 & -20 & 10 \\ 10 & 10 & 20 \end{bmatrix}$

4. $\begin{bmatrix} -20 & 10 & 10 \\ 10 & -20 & 10 \\ 10 & 10 & 20 \end{bmatrix}$

4.15 Key Answers to Four-Choice Questions

Question	1	2	3	4
1. (1)	×			
2. (4)				×
3. (2)		×		
4. (2)		×		
5. (2)		×		
6. (4)				×
7. (3)			×	
8. (3)			×	
9. (1)	×			
10. (1)	×			
11. (4)				×
12. (1)	×			
13. (3)			×	
14. (4)				×
15. (2)		×		
16. (4)				×
17. (3)			×	
18. (3)			×	
19. (3)			×	
20. (2)		×		
21. (4)				×
22. (3)			×	
23. (1)	×			
24. (2)		×		
25. (4)				×
26. (1)	×			
27. (1)	×			
28. (2)		×		
29. (1)	×			
30. (2)		×		
31. (3)			×	
32. (3)			×	
33. (4)				×
34. (1)	×			
35. (2)		×		
36. (3)			×	
37. (4)				×
38. (2)		×		
39. (4)				×

(Continued)

Question	1	2	3	4
40. (1)	×			
41. (3)			×	
42. (2)		×		
43. (1)	×			
44. (3)			×	
45. (3)			×	
46. (3)			×	
47. (4)				×
48. (2)		×		
49. (1)	×			
50. (2)		×		
51. (4)				×
52. (1)	×			
53. (2)		×		
54. (2)		×		
55. (1)	×			

4.16 Descriptive Answers to Four-Choice Questions

4.1. **Option 1 is correct.** Since we know (equation 4.46) and X_{ii} in the matrix of $[X]$ is the seen impedance from the ith bus, and since only ΔP_1 has a value, we don't need to calculate the matrix of $[B]$ and its inverse. As a result, we have:

From bus 1, the impedance is as follows: $X_{11} = 0.4 \parallel (0.2 + 0.2) = 0.2$

And we have: $\Delta P_1 = \dfrac{10}{100} \times \dfrac{65\,\text{MW}}{100\,\text{MW}} = 0.065,\ \Delta P_2 = 0$

$$\Rightarrow \Delta \delta_1 = X_{11}\Delta P_1 = 0.2 \times 0.065 = 0.013 \text{ rad}$$

4.2. **Option 4 is correct.** In this question, we need to calculate the matrix of $[B]$ and its inverse in relation to equation (4.46) achieved. We have:

$$B = \begin{bmatrix} 7.5 & -5 \\ -5 & 10 \end{bmatrix} \Rightarrow X = B^{-1} = \frac{1}{50} \begin{bmatrix} 10 & 5 \\ 5 & 7.5 \end{bmatrix}$$

$$\Delta P_1 = \frac{10}{100} \times \frac{65\ \text{MW}}{100\,\text{MW}} = 0.065,$$

$$\Delta P_2 = \frac{6.5}{100} \times \frac{-100\,\text{MW}}{100\,\text{MW}} = -0.065$$

In equation (4.43) or (4.46), P and/or ΔP represent the injection power of $(P_G - P_D)$ into the network.

$$\Rightarrow \begin{bmatrix} \Delta\delta_1 \\ \Delta\delta_2 \end{bmatrix} = \frac{1}{50} \begin{bmatrix} 10 & 5 \\ 5 & 7.5 \end{bmatrix} \begin{bmatrix} 0.065 \\ -0.065 \end{bmatrix}$$

$$\Rightarrow \Delta\delta_1 = \frac{1}{50}(10-5) \times 0.065 = 0.0065$$

4.3. **Option 2 is correct.** Using equation (4.55) we have:

$$1(1-0.4Q) \geq \frac{4 \times 4}{100} \Rightarrow 0.4Q \leq 1-0.16 \Rightarrow 0.4Q \leq 0.84 \Rightarrow Q \leq 2.1$$

4.4. **Option 2 is correct.** Using equation (4.50) we have:

$$V_2^2 = \frac{(1-2 \times 2 \times 0.1) \pm \sqrt{1(1-4 \times 2 \times 0.1) - 4 \times 0.01 \times 4}}{2}$$

$$= \frac{0.6 \pm \sqrt{1-0.8-0.16}}{2}$$

$$= \frac{0.6 \pm \sqrt{0.04}}{2} = \frac{0.6 \pm 0.2}{2} = \begin{cases} 0.4: & \text{Stable} \\ 0.2: \text{Unstable} \end{cases} \Rightarrow V_2 = \sqrt{0.4}$$

4.5. **Option 2 is correct.** According to the answer to question 4.4, $V_2 = \sqrt{0.2}$ is correct.

4.6. **Option 4 is correct.** Using equation (4.56) we have:

$$V_2 = \sqrt{\frac{1-2 \times 2 \times 0.1}{2}} = \sqrt{0.3}$$

If instead of Q of the load, P of the load was known, first Q is calculated using equation (4.55) and the equal condition, and then we implement (equation 4.56).

4.7. **Option 3 is correct.** Using equation (4.54) we have:

$$\hat{V}_2^{(1)} = \frac{-0.1(2+j2)}{1\angle 0} + 1 = -0.2 - j0.2 + 1 = 0.8 - j0.2$$

$$\Rightarrow V_2^{(1)} = \sqrt{0.64 + 0.04} = \sqrt{0.68}$$

4.8. **Option 3 is correct.** Based on the π model transformer with a tap in the relation of equation (4.62), we have (Figure 4.36):

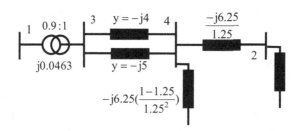

FIGURE 4.36
Answer system 4.8.

$$y_t = \frac{1}{j0.16} = -j6.25$$

The figure shows the admittances connected to bus 4 as follows:

$$y_{44} = -j5 - j4 - j\frac{6.25}{1.25} - j6.25\frac{-0.25}{1.25 \times 1.25}$$

$$= -j9 - j\frac{5 \times 1.25}{1.25} + j\frac{5 \times 1.25 \times 0.25}{1.25 \times 5 \times 0.25}$$

$$= -j9 - j5 + j1 = -j13 \Rightarrow |y_{44}| = 13$$

4.9. **Option 1 is correct.** In accordance with the relation of equation (4.40a), the matrix of $[B']$ is not related to the voltage control buses, only the reference bus has been removed. The admittance matrix for the system is:

$$Y_{bus} = \begin{bmatrix} -j7.5 & j5 & j2.5 \\ j5 & -j10 & j5 \\ j2.5 & j5 & -j7.5 \end{bmatrix} \Rightarrow B' = \begin{bmatrix} -10 & 5 \\ 5 & -7.5 \end{bmatrix}$$

$$\begin{bmatrix} \Delta P \\ V \end{bmatrix} = -[B'][\Delta\delta] \Rightarrow \begin{bmatrix} \dfrac{\Delta P_2}{V_2} \\ \dfrac{\Delta P_3}{V_3} \end{bmatrix} = -\begin{bmatrix} -10 & 5 \\ 5 & -7.5 \end{bmatrix}\begin{bmatrix} \Delta\delta_2 \\ \Delta\delta_3 \end{bmatrix}$$

$$\Rightarrow \begin{bmatrix} \Delta\delta_2 \\ \Delta\delta_3 \end{bmatrix} = \frac{1}{50} \begin{bmatrix} 7.5 & 5 \\ 5 & 10 \end{bmatrix} \begin{bmatrix} \dfrac{0.1}{1} \\ \dfrac{0.105}{1.05} \end{bmatrix}$$

$$\Rightarrow \Delta\delta_2 = \frac{1}{50}(7.5 \times 0.1 + 5 \times 0.1) = 0.025$$

4.10. **Option 1 is correct.** According to the relation of equation (4.40b) and the voltage control of bus 3, the matrix of $[B'']$ is obtained from the calculated admittance matrix in answer 4.9 as follows:

$$[B''] = -10 \Rightarrow \frac{\Delta Q_2}{V_2} = 10\Delta V_2 \Rightarrow \frac{1.01}{1.01} = 10\Delta V_2 \Rightarrow \Delta V_2 = 0.1$$

4.11. **Option 4 is correct.** DC LF and FDLF load flow differ in that $\Delta\delta$ is calculated according to the relation of $\Delta P/V$ (see equations 4.40a and 4.43). According to equation (4.43) and the calculated admittance matrix in answer 4.9, we have:

$$B = -B' = \begin{bmatrix} 10 & -5 \\ -5 & 7.5 \end{bmatrix} \Rightarrow \begin{bmatrix} \Delta P_2 \\ \Delta P_3 \end{bmatrix} = \begin{bmatrix} 10 & -5 \\ -5 & 7.5 \end{bmatrix} \begin{bmatrix} \Delta\delta_2 \\ \Delta\delta_3 \end{bmatrix}$$

$$\Rightarrow \begin{bmatrix} \Delta\delta_2 \\ \Delta\delta_3 \end{bmatrix} = \frac{1}{50} \begin{bmatrix} 7.5 & 5 \\ 5 & 10 \end{bmatrix} \begin{bmatrix} 0.1 \\ 0.105 \end{bmatrix}$$

$$\Rightarrow \Delta\delta_2 = \frac{1}{50}(7.5 \times 0.1 + 5 \times 0.105)$$

$$= \frac{0.75 + 0.525}{50} = \frac{1.275}{50} = \frac{2 \times 1.275}{100} = 0.0255$$

4.12. **Option 1 is correct.** Inverting the matrix is a simple solution. We have:

$$\text{Det}(Y_{bus}) = j(-3(24 - 9) - 1(-6 - 6) + 2(3 + 8)) = -11j$$

$$\Rightarrow Z_{11} = \frac{1}{-11j}((-4 \times -6) - 3 \times 3) = j\frac{15}{11}$$

4.13. **Option 3 is correct.** We must calculate Z_{11} and Z_{12} using equation (4.63). Like the answer to question 4.12, we have:

$$Z_{11} = j\frac{15}{11}, \quad Z_{12} = \frac{-1}{-11j}((1 \times -6) - 2 \times 3) = j\frac{12}{11}$$

$$V_{22} = \left| \hat{V}_2^o - Z_{21} \times \frac{\hat{V}_1^o}{Z_f + Z_{11}} \right| = \left| 1.2 - j\frac{12}{11} \times \frac{1.1}{j\frac{9}{11} + j\frac{15}{11}} \right| = |1.2 - 0.5 \times 1.1| = 0.65$$

4.14. **Option 4 is correct.** When one bus is removed from the system, the Thevenin impedance of the other buses does not change, but the admittance matrix does. So we have:

$$\text{Det}\left(Y_{\text{bus}}\right) = j\left(-3(24-9) - 1(-6-6) + 2(3+8)\right) = -11j$$

$$\Rightarrow Z_{22} = \frac{1}{-11j}\left((-3 \times -6) - 2 \times 2\right) = j\frac{14}{11}$$

4.15. **Option 2 is correct.** Note $\left(y_{11} \neq \dfrac{1}{Z_{11}}\right)$. Using equations (4.74) and (4.75) we have:

$$\begin{bmatrix} I_A \\ I_X \end{bmatrix} = \begin{bmatrix} K & L \\ L^T & M \end{bmatrix} \begin{bmatrix} V_A \\ V_X \end{bmatrix},$$

$$K = j\begin{bmatrix} -3 & 1 \\ 1 & -4 \end{bmatrix}, \quad L = j\begin{bmatrix} 2 \\ 3 \end{bmatrix}, \quad L^T = j\begin{bmatrix} 2 & 3 \end{bmatrix},$$

$$M = -j6 \Rightarrow [Y_{\text{red}}] = K - LM^{-1}L^T$$

$$\Rightarrow [Y_{\text{red}}] = j\left(\begin{bmatrix} -3 & 1 \\ 1 & -4 \end{bmatrix} - \frac{1}{-6}\begin{bmatrix} 2 \\ 3 \end{bmatrix} \times \begin{bmatrix} 2 & 3 \end{bmatrix}\right)$$

$$= j\begin{bmatrix} -3 + \dfrac{4}{6} & 1 + \dfrac{6}{6} \\ 1 + \dfrac{6}{6} & -4 + \dfrac{9}{6} \end{bmatrix} = j\begin{bmatrix} -\dfrac{7}{3} & 2 \\ 2 & -\dfrac{5}{2} \end{bmatrix}$$

4.16. **Option 4 is correct.** Using equation (4.70) with $k=1$ and $h=2$ we have:

$$I_f = \left| \frac{\hat{V}_1^o - \hat{V}_2^o}{Z_{22} + Z_{11} + Z_f - 2Z_{12}} \right| = \left| \frac{1.1 - 1.2}{j0.3 + j0.2 - j0.15 - 2 \times j0.15} \right| = \frac{0.1}{0.05} = 2$$

4.17. **Option 3 is correct.** Using equations (4.70) and (4.71) we have:

$$\hat{V}_3^f = \hat{V}_3^o - (Z_{31} - Z_{32}) \times \frac{\hat{V}_1^o - \hat{V}_2^o}{Z_{22} + Z_{11} + Z_f - 2Z_{12}}$$

$$= 1.1 - j(0.1 - 0.15) \times \frac{1.1 - 1.2}{j(0.3 + 0.2 - 0.15 - 2 \times 0.15)}$$

$$= 1.1 - 0.1 = 1$$

4.18. **Option 3 is correct.** Using equation (4.73) we have:

$$\frac{\Delta V_2}{\Delta V_3} = \left| \frac{Z_{21} - Z_{22}}{Z_{31} - Z_{32}} \right| = \left| \frac{0.15 - 0.3}{0.1 - 0.15} \right| = \frac{0.15}{0.05} = 3$$

4.19. **Option 3 is correct.** We have:

1. $P_{L\,ac} = 3R_L I_L^2, \quad P_{L\,dc} = 2R_L I_d^2 \Rightarrow \frac{P_{L\,ac}}{P_{L\,dc}} = \frac{3}{2}\left(\frac{I_L}{I_d}\right)^2$

2. $\left. \begin{array}{l} P_{ac} = 3V_{ph}I_L \\ P_{dc} = 2V_d I_d \end{array} \right\} \Rightarrow 3\frac{V_{max}}{\sqrt{2}}I_L = 2V_d I_d \xRightarrow{V_{max}=V_d} \frac{I_L}{I_d} = \frac{2\sqrt{2}}{3}$

$$\xRightarrow{(1),(2)} \frac{P_{L\,ac}}{P_{L\,dc}} = \frac{3}{2}\left(\frac{2\sqrt{2}}{3}\right)^2 = \frac{3}{2}\left(\frac{8}{9}\right) = \frac{4}{3}$$

4.20. **Option 2 is correct.** Since the isolation level is proportional to the maximum voltage, we have:

1. $P_{L\,ac} = P_{L\,dc} \Rightarrow 3R_L I_L^2 = 2R_L I_d^2 \Rightarrow I_d = \sqrt{\frac{3}{2}}\,I_L$

$$P_{ac} = P_{dc} \Rightarrow 3V_{ph}I_L = 2V_d I_d \xRightarrow{(1)} 3\frac{V_{max}}{\sqrt{2}}I_L = 2V_d\sqrt{\frac{3}{2}}\,I_L$$

$$\Rightarrow 3V_{max} = 2V_d\sqrt{3} \Rightarrow \frac{V_d}{V_{max}} = \frac{3}{2\sqrt{3}} = \frac{\sqrt{3}}{2}$$

4.21. **Option 4 is correct.** As the base voltage and power for the two systems of connection - AC and DC - must be the same, we have:

$$\left. \begin{array}{l} V_{b\,dc} = V_{b\,ac} = V_b \\ P_{b\,dc} = P_{b\,ac} = P_b \end{array} \right\} \Rightarrow \underbrace{\sqrt{3}\,V_b I_{b\,ac}}_{P_{b\,ac}} = \underbrace{V_b I_{b\,dc}}_{P_{b\,dc}} \Rightarrow \frac{I_{b\,dc}}{I_{b\,ac}} = \sqrt{3}$$

4.22. **Option 3 is correct.** Given the answer to question 4.19, with the difference that the isolation level ratio is not one and we have $\dfrac{V_d}{V_{max}} = x$, then we have:

$$\Rightarrow \frac{P_{L\,ac}}{P_{L\,dc}} = \frac{4}{3}x^2 \Rightarrow P_{L\,dc} = \frac{3}{4\times0.9^2}\times81 = \frac{3\times81}{4\times0.81} = 75$$

4.23. **Option 1 is correct.** Based on the answer to question 4.21 and equality of real powers, we have:

$$V_d I_d = \sqrt{3}\,V_{ac}I_{ac}\cos(\varphi)$$

$$\Rightarrow \left(V_{d\,p.u.}V_b\right)\left(I_{d\,p.u.}\sqrt{3}\,I_{bac}\right) = \sqrt{3}\left(V_{ac\,p.u.}V_b\right)\left(I_{ac\,p.u.}I_{b\,ac}\right)\cos(\varphi)$$

$$\Rightarrow V_{d\,p.u.}I_{d\,p.u.} = V_{ac\,p.u.}I_{ac\,p.u.}\cos(\varphi) = P_{ac\,p.u.} = |S|\cos(\varphi)$$

$$\Rightarrow 1.2\times I_{dc\,p.u.} = 0.9\times0.8$$

$$\Rightarrow I_{dc\,p.u.} = \frac{0.72}{1.2} = 0.6$$

4.24. **Option 2 is correct.** Based on equation (4.76), since the line losses are determined by line current, which is also determined by generator 1, the losses are not determined by generator 2. Therefore, we have:

$$P_L = B_{11}P_1^2 \Rightarrow B_{11} = \frac{5}{100^2} = 0.0005 \Rightarrow P_{L2} = 0.0005\times80^2 = 3.2$$

4.25. **Option 4 is correct.** Using equation (4.78) we have:

$$P_L = B_{11}P_1^2 + B_{22}P_2^2 + 2B_{12}P_1P_2 \Rightarrow \frac{\Delta P_L}{\Delta P_1} \approx \frac{\partial P_L}{\partial P_1} = 2B_{11}P_1 + 2B_{12}P_2$$

$$B_{11} = \frac{R_1}{\left(V_1\cos\varphi_1\right)^2} + \frac{R_3}{\left(V_3\cos\varphi_3\right)^2} \Rightarrow B_{11} = \frac{0.2}{\left(1\times1\right)^2} + \frac{0.81}{\left(1\times0.9\right)^2} = 1.2$$

$$B_{12} = \frac{R_3}{\left(V_3\cos\varphi_3\right)^2} \Rightarrow B_{12} = \frac{0.81}{\left(1\times0.9\right)^2} = 1$$

$$\Rightarrow \frac{\Delta P_L}{2\%} \approx 2\times1.2\times\frac{5}{6} + 2\times1\times0.8 = 3.6 \Rightarrow \Delta P_L = 7.2\%$$

4.26. **Option 1 is correct.** Sparsity is defined in relation to equation (4.48) as follows:

It is shown that $\dfrac{n(n-1)}{2}$ is equal to the number of off-diagonal arrays of the admittance matrix arrays when n is the number of buses and l is the number of lines in the system.

Therefore, the total number of zeros in the admittance matrix arrays is:

$$Z = 2 \times \left(\frac{n(n-1)}{2} - l \right) = n^2 - n - 2l$$

$$\Rightarrow S = \frac{Z}{n^2} = \frac{10^2 - 10 - 2 \times 40}{10^2} = 0.1$$

4.27. **Option 1 is correct.** A simple example of sparsity can be found in the summary of the relations. An admittance matrix in the normal case requires the memory of:

$$n + \frac{n(n-1)}{2} = \frac{n(n+1)}{2} = \frac{10 \times 11}{2} = 55$$

There are n main diameters and $\dfrac{n(n-1)}{2}$ non-diagonals in the admittance matrix. If we use Sparsity for programming, we need the memory of $3 \times l$ for the store of I_R, I_C, and Z_{offd}, and the memory of n for the store of Z_{diag}, so for the Sparsity case we have:

$$n + (3 \times l) = 10 + (3 \times 40) = 130$$

4.28. **Option 2 is correct.** According to the answer to question 4.27, we have:

$$3l + n < \frac{n(n+1)}{2} \Rightarrow 3 \times l + 10 < \frac{10 \times 11}{2} \Rightarrow 3l < 45 \Rightarrow l < 15$$

4.29. **Option 1 is correct.** Using equation (4.82) we have:

$$\Delta U\% = \frac{200 \sum\limits_{i=1}^{n} l_i P_i}{\sigma A U^2} = \frac{200(30 \times 30k + 70 \times 10k + 90 \times 30k)}{35 \times \dfrac{1000}{7} \times 200^2}$$

$$= \frac{(900 + 700 + 2700) \times 1000}{5000 \times 200} = \frac{4300k}{10^6} = 4.3\%$$

4.30. **Option 2 is correct.** Using equation (4.95), we calculate the updated lengths based on the largest cross-sectional area.

$$l_1' = l_1, \quad l_2' = l_2\frac{A'}{A_2} = 40 \times \frac{150}{50} = 120, \quad l_3' = l_3\frac{A'}{A_{32}} = 20 \times \frac{150}{25} = 120$$

$$\overset{(4.82)}{\Rightarrow} \Delta U\% = \frac{200(30 \times 30k + 150 \times 10k + 270 \times 30k)}{35 \times 150 \times 200^2}$$

$$= \frac{900k + 1500k + 8100k}{35 \times 150 \times 200} = \frac{10,500k}{35 \times 3 \times 10^4} = \frac{3 \times 35 \times 10^5}{35 \times 3 \times 10^4} = 10\%$$

4.31. **Option 3 is correct.** The shape of the question should be converted into equation (4.86) as follows (Figure 4.37):

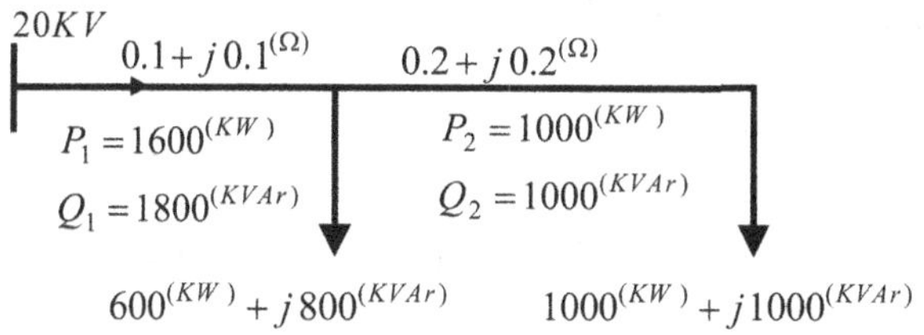

FIGURE 4.37
Answer system 4.31.

$$\Delta U = \frac{0.1 \times 1600 + 0.1 \times 1800 + 0.2 \times 1000 + 0.2 \times 1000}{20k}$$

$$= \frac{740k}{20k} = 37 \text{ V}$$

4.32. **Option 3 is correct.** Using equation (4.89) we have:

$$I_A = \frac{50l + 50 \times 2l + 50 \times 3l}{4l} = 75$$

Due to the symmetry, we can also arrive at the same conclusion, so we have:

$$I_A = I_B = \frac{3 \times 50}{2} = 75$$

4.33. **Option 4 is correct.** In the same way as in problem 4.32, both symmetry and the relation (4.89) lead to the following:

$$I_A = \frac{I \times l + I \times 2l + \cdots + I \times nl}{(n+1)l} = \frac{I\frac{n(n+1)}{2}}{n+1} = \frac{nI}{2}$$

4.34. Option 1 is correct. Using equation (4.86) we have:

$$\Delta U = \frac{1}{U}\sum r_i P_i = \frac{rlP + (2rlP) + \cdots + (nrlP)}{U} = \frac{rlP(1 + 2 + \cdots n)}{U}$$

$$= \frac{rlPn(n+1)}{2U}$$

4.35. Option 2 is correct. Using equation (4.89), or the answer to the question of 4.33, and/or using symmetry, we have $P_A = P_B = \frac{nP}{2}$. If n is even, the load numbers $\frac{n}{2}$ and $\frac{n}{2} + 1$ are the deep points.

$$\Delta U = \frac{1}{U}\sum_{i=1}^{\frac{n}{2}} r_i l P_i = \frac{rl}{U}\left(\frac{nP}{2} + \left(\frac{nP}{2} - P\right) + \cdots + \left(\frac{nP}{2} - \left(\frac{n}{2} - 1\right)P\right)\right)$$

$$= \frac{rlP}{U}\left[\underbrace{\frac{n}{2} + \cdots + \frac{n}{2}}_{\frac{n}{2}} - \left(1 + 2 + \cdots + \left(\frac{n}{2} - 1\right)\right)\right] = \frac{rlP}{U}\left[\frac{n^2}{4} - \frac{\left(\frac{n}{2} - 1\right)\frac{n}{2}}{2}\right]$$

$$= \frac{rlP}{U}\left(\frac{n^2}{4} - \frac{(n-2)n}{8}\right) = \frac{rlP(n^2 + 2n)}{8U}$$

4.36. Option 3 is correct. Using equation (4.89), or the answer to the question of 4.33, and/or using symmetry, we have $P_A = P_B = \frac{nP}{2}$. If n is odd, the load number of $\frac{n+1}{2}$ is the deep point.

$$\Delta U = \frac{1}{U}\sum_{i=1}^{\frac{n+1}{2}} r_i l P_i = \frac{rl}{U}\left[\frac{nP}{2} + \left(\frac{nP}{2} - P\right) + \cdots + \left(\frac{nP}{2} - \left(\frac{n-1}{2}\right)P\right)\right]$$

$$= \frac{rlP}{U}\left[\underbrace{\frac{n}{2} + \cdots + \frac{n}{2}}_{\frac{n+1}{2}} - \left(1 + 2 + \cdots + \left(\frac{n-1}{2}\right)\right)\right] = \frac{rlP}{U}\left[\frac{n(n+1)}{2 \cdot 2} - \frac{\left(\frac{n-1}{2}\right)\left(\frac{n+1}{2}\right)}{2}\right]$$

$$= \frac{rlP}{U}\left(\frac{n^2 + n}{4} - \frac{n^2 - 1}{8}\right) = \frac{rlP}{U}\left(\frac{2n^2 + 2n - n^2 + 1}{8}\right) = \frac{rlP}{U}\left(\frac{(n+1)^2}{8}\right)$$

4.37. **Option 4 is correct.** First, we must find the deep point. Assume we are starting from the side of B. Using equation (4.89) we find the deep point and using equation (4.90) the percentage of the voltage drop is calculable (Figure 4.38):

FIGURE 4.38
Answer system 4.37.

$$I_B = \frac{10 \times 20 + 20 \times 40 + 30 \times 50 + 30 \times 60}{20 + 20 + 10 + 10 + 40}$$

$$= \frac{200 + 800 + 1500 + 1800}{100} = 43$$

$$\Rightarrow \Delta U\% = \frac{\overset{\text{deep}}{\underset{i=1}{\sum} l_i I_i}}{\sigma A U} \times 100 = 5 \Rightarrow \frac{43 \times 40 + 13 \times 10}{200 \times 50 \times A} \times 100 = 5$$

$$\Rightarrow A = \frac{1720 + 130}{2 \times 50 \times 5} = \frac{1850}{500} = 3.7$$

4.38. **Option 2 is correct.** Similarly to the problem of 4.37, we first found the deep point, then calculated the voltage drop (see relations of 4.93 and 4.94) (Figure 4.39).

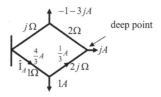

FIGURE 4.39
Answer system 4.38.

$$\hat{I}_A = \frac{j(-1-3j) + j(2+j) + 1(2+3j)}{j + 2 + 2j + 1}$$

$$= \frac{-j + 3 + 2j - 1 + 2 + 3j}{3(1+j)} = \frac{4j + 4}{3(1+j)} = \frac{4}{3}$$

$$\Rightarrow \Delta U = \frac{4}{3} \times 1 + \frac{1}{3} \times 2j \Rightarrow |\Delta U| = \sqrt{\frac{16+4}{9}} \Rightarrow |\Delta U| = \frac{2\sqrt{5}}{3}$$

4.39. **Option 4 is correct.** Using equations (4.11) and (4.12) we have:

$$a_{ij} = -B_{ij}f_j, \quad b_{ij} = B_{ij}e_j$$

$$\Rightarrow P_i = e_i \sum_{j=1}^{n} a_{ij} + f_i \sum_{j=1}^{n} b_{ij} = e_i \sum_{j=1}^{n} \left(-B_{ij}f_j\right) + f_i \sum_{j=1}^{n} \left(B_{ij}e_j\right)$$

$$\Rightarrow \frac{\partial P_i}{\partial e_i} = -\sum_{j=1}^{n} B_{ij}f_j + f_i B_{ii} = -\sum_{j=1\neq i}^{n} B_{ij}f_j$$

4.40. **Option 1 is correct.** Subtracting equations (4.17) and (4.18) gives us:

$$\Delta \hat{V}_i^{(r+1)} = \sum_{j=1}^{i-1} \frac{Y_{ij}}{Y_{ii}} \left(\hat{V}_j^{(r+1)} - \hat{V}_j^{(r)}\right)$$

$$\Rightarrow \Delta \hat{V}_3^{(2)} = \sum_{j=1}^{2} \frac{Y_{3j}}{Y_{33}} \left(\hat{V}_j^{(2)} - \hat{V}_j^{(1)}\right) = \frac{Y_{31}}{Y_{33}} \left(\hat{V}_1^{(2)} - \hat{V}_1^{(1)}\right) + \frac{Y_{32}}{Y_{33}} \left(\hat{V}_2^{(2)} - \hat{V}_2^{(1)}\right)$$

$$= 0 + \frac{j3}{-j6}(1.2 - 1.1) = -0.05$$

4.41. **Option 3 is correct.** By using equation (4.28), we obtain the approximate relation that is also true at the working point:

$$J_{1_{22}} = \frac{\partial f_{P2}}{\partial \delta_2} = -Q_2 - V_2^2 B_{22}, \quad Y_{22} = -j5 - j5 = -j10 \Rightarrow B_{22} = -10$$

$$Q_2 = Q_{G2} - Q_{D2} = 0 - \frac{250}{100} = -2.5 \Rightarrow J_{1_{22}} = 2.5 - 1 \times (-10) = 12.5$$

4.42. **Option 2 is correct.** In the use of equations (4.31) and (4.32) and this issue that the row and column corresponding to the reference bus (bus 1) have been removed from matrices J_1 until J_4, we have:

$$J_{1_{22}} = -Q_2 - V_2^2 B_{22} = \left(\text{The first row and column array of the matrix of } J_1\right)$$

$$J_{3_{23}} = -V_2 V_3 |Y_{23}| \cos(\delta_2 - \delta_3 - \gamma_{23})$$

$$= \left(\text{The first row and the second column array of the matrix of } J_3\right)$$

$$Y_{22} = -j5 - j5 = -j10 \Rightarrow B_{22} = -10,$$

$$Q_2 = Q_{G2} - Q_{D2} = 0 - \frac{250}{100} = -2.5 \Rightarrow J_{1_{22}} = 2.5 - 1 \times (-10) = 12.5$$

$$Y_{23} = -j5 \Rightarrow J_{3_{23}} = -1 \times 1 \times 5\cos(30 - 60 - (-90)) = -5\cos(60) = -2.5$$

$$\Rightarrow \frac{J_{1_{22}}}{J_{3_{23}}} = \frac{12.5}{-2.5} = -5$$

4.43. **Option 1 is correct.** Using equation (4.35) we have:

$$\|\Delta X\|_1 = \sum_{i=1}^{n} |\Delta X_i| = 0.15 + 0.2 + 0.1 + 0.05 = 0.5$$

$$\|\Delta X\|_\infty = \max|\Delta X_i| = 0.2 \Rightarrow \frac{\|\Delta X\|_1}{\|\Delta X\|_\infty} = \frac{0.5}{0.2} = 2.5$$

4.44. **Option 3 is correct.**

$$Y_{bus} = \begin{bmatrix} -j2 & j \\ j & -j3 \end{bmatrix} \Rightarrow Z_{bus} = Y_{bus}^{-1} \Rightarrow \text{Det}(Z_{bus})$$

$$= \frac{1}{\text{Det}(Y_{bus})} = \frac{1}{-6+1} = -0.2$$

4.45. **Option 3 is correct.** Using equations (4.43) and (4.45) we have:

$$P_2 = ?, P_3 = 0 - 1 = -1, \delta_2 = ?, \delta_3 = -12 \times \frac{\pi}{180} = -12 \times \frac{3}{180} = -0.2$$

$$Y_{bus} = \begin{bmatrix} ? & ? & ? \\ ? & -j2 & j2 \\ ? & j2 & -j4 \end{bmatrix} \Rightarrow B' = \begin{bmatrix} -2 & 2 \\ 2 & -2 \end{bmatrix} \Rightarrow B = \begin{bmatrix} 2 & -2 \\ -2 & 2 \end{bmatrix}$$

$$\Rightarrow \begin{bmatrix} P_2 \\ -1 \end{bmatrix} = \begin{bmatrix} 2 & -2 \\ -2 & 2 \end{bmatrix} \begin{bmatrix} \delta_2 \\ -0.2 \end{bmatrix} \Rightarrow -1 = -2\delta_2 + 4(-0.2)$$

$$\Rightarrow \delta_2 = \frac{-1 + 0.8}{-2} = 0.1$$

$$\Rightarrow P_2 = 2(0.1) - 2(-0.2) = 0.6$$

4.46. **Option 3 is correct.** Using equation (4.63) we have:

$$\hat{V}_4^f = \hat{V}_4^o - Z_{44} \frac{\hat{V}_4^o}{Z_f + Z_{44}} = 0.9 - j0.4 \frac{0.9}{-j3.4 + j0.4}$$

$$= 0.9 - \frac{j0.4 \times 0.9}{-j3} = 0.9 + 0.4 \times 0.3 = 1.02$$

4.47. **Option 4 is correct.** As the economic dispatch depends on the derivative of costs, and since $(\beta_1 > \beta_2)$ $(50 > 40)$ and $(\gamma_1 > \gamma_2)$ $(0.0075 > 0.005)$ is greater for generator 1 than for generator 2, as a result, P_{G1} must be greater than P_{G2}, and only option 4 is correct $(800 > 200)$.

4.48. **Option 2 is correct.** First, we find the deep point in Figure 4.40.

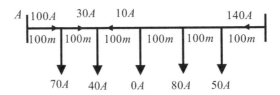

FIGURE 4.40
Answer system 4.48.

According to Equation (4.94), the voltage drop from A is equal to:

$$\Delta U = 100 \times 100 \times 10^{-4} + 100 \times 30 \times 10^{-4} = 1 + 0.3 = 1.3$$

4.49. **Option 1 is correct.** When one of the lines is disconnected from bus 1 and connected to earth, Y_{11} becomes low. Therefore options 1 and 4 are correct, but Y_{22} remains unchanged, so option 1 is correct.

4.50. **Option 2 is correct.** The resistance of each part becomes $R/5$ if the resistance of the entire line is R. By using equation (4.93) or KVL, we have:

$$I_a = \frac{230 - 220 + 100 \times \dfrac{R}{5} + 100 \times \dfrac{2R}{5} + 100 \times \dfrac{3R}{5} + 100 \times \dfrac{4R}{5}}{R}$$

$$= \frac{10}{0.2} + \frac{100}{5} + \frac{200}{5} + \frac{300}{5} + \frac{400}{5} = 50 + 20 + 40 + 60 + 80 = 250$$

$$\Rightarrow V_d = 230 - 250 \frac{R}{5} - 150 \frac{R}{5} - 50 \frac{R}{5} = 230 - 450 \frac{0.2}{5} = 230 - 18 = 212$$

4.51. **Option 4 is correct.** The admittance matrix between buses 1 and 3 has been used and then we remove bus 3.

$$\begin{bmatrix} \hat{I}_1 \\ \hat{I}_3 \end{bmatrix} = \begin{bmatrix} Y_1 + Y_3 & -Y_3 \\ -Y_3 & Y_2 + Y_3 \end{bmatrix} \begin{bmatrix} \hat{V}_1 \\ \hat{V}_3 \end{bmatrix}$$

We have the following transformers with taps: $\dfrac{\hat{V}_3}{\hat{V}_2} = a$, $\dfrac{\hat{I}_2}{\hat{I}_3} = a^*$

We find the following difference between the above relation and the relation of equation (4.60) in the direction of the current $\hat{I}_3$ in the figure of the question.

$$\begin{bmatrix} \hat{I}_1 \\ \dfrac{\hat{I}_2}{a^*} \end{bmatrix} = \begin{bmatrix} Y_1 + Y_3 & -Y_3 \\ -Y_3 & Y_2 + Y_3 \end{bmatrix} \begin{bmatrix} \hat{V}_1 \\ a \times \hat{V}_2 \end{bmatrix} \Rightarrow \begin{bmatrix} \hat{I}_1 \\ \hat{I}_2 \end{bmatrix}$$

$$= \begin{bmatrix} Y_1 + Y_3 & -aY_3 \\ -a^*Y_3 & |a|^2 (Y_2 + Y_3) \end{bmatrix} \begin{bmatrix} \hat{V}_1 \\ \hat{V}_2 \end{bmatrix}$$

The answer is found by multiplying the second row in (a^*) and the second column in (a).

4.52. **Option 1 is correct.** By using equations (4.3) and (4.8) we have:

$$S_i = \hat{V}_i \cdot \hat{I}_i^* = \hat{V}_i \left(\sum_{j=1}^{n} Y_{ij} \hat{V}_j \right)^* \Rightarrow \frac{S_i^*}{\hat{V}_i^*} = \sum_{j=1}^{n} Y_{ij} \hat{V}_j$$

$$\Rightarrow \frac{P_i - jQ_i}{V_i \times e^{-j\delta_i}} = \frac{(P_i - jQ_i) \times e^{j\delta_i}}{V_i} = \sum_{j=1}^{n} Y_{ij} \hat{V}_j$$

$$\Rightarrow \frac{(P_i - jQ_i)(\cos\delta_i + j\sin\delta_i)}{V_i} = \sum_{j=1}^{n} V_j (\cos\delta_j + j\sin\delta_j)(G_{ij} + jB_{ij})$$

$$\overset{\text{Imaginary}}{\Rightarrow} \frac{Q_i \cos\delta_i - P_i \sin\delta_i}{V_i} = -\sum_{j=1}^{n} V_j (B_{ij}\cos\delta_j + G_{ij}\sin\delta_j)$$

$$\overset{\text{real}}{\Rightarrow} \frac{(P_i \cos\delta_i + Q_i \sin\delta_i)}{V_i} = \sum_{j=1}^{n} V_j (G_{ij}\cos\delta_j - B_{ij}\sin\delta_j)$$

4.53. **Option 2 is correct.** Assuming equation (4.64), in this question, $i=3$ is the desired bus and $k=2$ is the short-circuited bus, so we have:

$$\Delta V_3 = \frac{X_{32} \times V_2}{X_{22} + X_f} = \frac{0.15 \times 1.2}{0.3 + 2.7} = 0.06$$

4.54. **Option 2 is correct.** According to the matrix, the admittance of each line is $-j10$, with lines 2–4 and 1–3 being outages. Thus, the admittance for all of Y_{ii} $\left(Y_{ii} = \sum_{j=1}^{n} Y_{Sij} \right)$ is low, hence $j10$ is added to the main diagonal of the admittance matrix. Also, $-j10$ must be added to the off-diagonal ($Y_{ij} = -Y_{Sij}$) so option 2 is correct.

4.55. **Option 1 is correct.** First of all, there is no AC system visible in the LF admittance matrix of the DC system (buses 1–3). Second, there is no imaginary part in this matrix. In this system, buses 2 and 3 are symmetrical, so Y_{22} and Y_{33} must be equal (Options 3 and 4 are incorrect).

4.17 Two-Choice Questions (Yes – No) – 59 Questions

1. Transient stability analysis and planning require an LF analysis.
2. The admittance matrix of a real power system is sparse.
3. Gauss-Seidel requires a large number of repetitions to achieve the desired accuracy, and convergence cannot be guaranteed.
4. In the Gauss-Seidel algorithm, a large acceleration coefficient reduces the number of repetitions.
5. By using the Gauss-Jacobi method, the updated values of variables obtained from the previous equations are immediately used in solving the next equation.
6. Acceleration coefficients can be used to increase convergence velocity.
7. Newton-Raphson was founded based on the Taylor series.
8. Newton-Raphson becomes convergent for each initial estimate.
9. Using Newton's method to solve a system of nonlinear equations reduces the problem of solving a system of linear equations.
10. Solving the LF involves determining the magnitude and angle of the voltages in the load buses, as well as the active and reactive power in each line.

11. In general, there are four types of power system buses.

12. Due to network losses, the voltage control bus compensates for the difference between the programmed loads and the generated power.

13. The magnitude and angle of the slack bus voltage are known.

14. The reactive power of the PV and PQ buses is known.

15. There is a generator on the voltage-regulating bus.

16. During normal operation, the bus voltage magnitude is 1 p.u. or close to the magnitude of the slack bus voltage.

17. During normal operation, the bus voltage magnitude is bigger than the magnitude of the slack bus voltage.

18. During normal operation, the bus voltage angle is significantly larger than the angle of the slack bus voltage.

19. The Gauss-Seidel method estimates the initial voltage as $1+j0$ for unknown voltages.

20. Voltage regulators and tap-changing transformers can be used to control active and reactive power.

21. In tap-changing transformers, when the tap is not real, we can represent the transformer with the following asymmetric π model.

22. Compared to Newton's method, the Gauss-Seidel method has a lower probability of divergence in a poorly structured system.

23. Gauss-Seidel is more practical and effective in large power systems.

24. In Newton's method, there is only one equation for each bus with controlled voltage.

25. According to the Jacobian matrix, the voltage angle and magnitude of small changes are linearized in terms of active and reactive power changes.

26. If the m bus in the n bus power system is controlled, the Jacobian matrix dimensions of that are $(2n-2-m)(2n-2-m)$.

27. J_3 has the dimensions $(n-1-m)$ $(n-1-m)$ in the Jacobian matrix of $\begin{bmatrix} J_1 & J_2 \\ J_3 & J_4 \end{bmatrix}$

28. In the fast decoupled load flow (FDLF) method, J_2 and J_3 from the Jacobian matrix are considered zero.

29. Newton's method takes more time for each repetition than the fast decoupled method.

30. In the contingency analysis, FDLF is very helpful.

31. The FDLF method converges in higher iterations for systems with a high X-to-R ratio.

32. LF includes active power limits.

33. Bus bars (PQ) are unproductive buses.

34. A voltage control bus is any bus whose voltage amplitude is kept constant.

35. There is an Ng voltage control bus (other than the slack bus) in an N-bus system, so there are 2N-Ng-2 equations to solve.

36. The LF problem is solved by considering the transmission lines with the π model.

37. In the node power, the effect of the reactive power compensators is apparent.

38. An actual power system has a symmetric admittance matrix.

39. If the transmission line is congested, Gauss-Seidel convergence becomes difficult.

40. With the Newton-Raphson method, computation time increases linearly with system size.

41. Newton-Raphson is suitable for solving very large systems that require an accurate solution.

42. Essentially, FDLF approximates Newton-Raphson's method.

43. For systems with a high ratio of X to R, a fast decoupled method is suitable.

44. It is generally compressed when the system size is reduced.

45. In the admittance matrix, Y_{ij} is equal to the line series admittance between bus i and j.

46. $S_D + S_G$ determines the injection power to the bus.

47. There is a known amount of reactive power produced on the voltage control bus.

48. If $Q > Q_{max}$, then the voltage control bus (PV) converts to the load bus (PQ).

49. P and Q are unknown on the reference bus.

50. The reference bus acts as the frequency regulator for the LF.

51. The Gauss method is simpler than Newton's.

52. The accuracy of the calculations is increased when the acceleration coefficient is used.

53. Gauss-Seidel is faster than Gauss-Jacobi.

54: FDLF is 100 times faster than Newton-Raphson.

55. FDLF has low accuracy.

56. The Jacobin matrix is usually inverted in LF calculations.

57. With the Newton-Raphson method, the number of repetitions is determined by the number of buses.

58. Newton's method is more challenging than Gauss's due to the large number of buses.

59. When all repetition steps are taken from the acceleration coefficient, the equation oscillates.

4.18 Key Answers to Two-Choice Questions

1. Yes
2. Yes
3. Yes
4. No
5. Yes
6. Yes
7. Yes
8. No
9. Yes
10. Yes
11. No
12. No
13. Yes
14. No
15. No
16. Yes
17. No
18. No
19. Yes
20. Yes
21. No
22. No
23. No
24. Yes
25. Yes
26. Yes
27. No
28. Yes

29. Yes
30. Yes
31. No
32. No
33. No
34. Yes
35. Yes
36. Yes
37. Yes
38. No
39. Yes
40. Yes
41. Yes
42. Yes
43. No
44. Yes
45. No
46. No
47. No
48. Yes
49. Yes
50. Yes
51. Yes
52. No
53. Yes
54. Yes
55. Yes
56. No
57. No
58. No
59. Yes

Further Reading

Baydokhty, M.E., M. Eidiani, H. Zeynal, H. Torkamani, H. Mortazavi, "Efficient generator tripping approach with minimum generation curtailment based on Fuzzy system rotor angle prediction", *Przeglad Elektrotechniczny*, 2012, 88(9A), pp. 266–271.

Eidiani, M., "A new hybrid method to assess available transfer capability in AC–DC networks using the wind power plant interconnection," *IEEE Systems Journal*, 2022, doi: 10.1109/JSYST.2022.3181099.

Eidiani, M., "A new load flow method to assess the static available transfer capability," *Journal of Electrical Engineering and Technology*, 2022, 17(5), pp. 2693–2701, doi: 10.1007/s42835-022-01105-3.

Eidiani, M., "A new method for assessment of voltage stability in transmission and distribution networks", *International Review of Electrical Engineering*, 2010, 5(1), pp. 234–240.

Eidiani, M., "A reliable and efficient holomorphic approach to evaluate dynamic available transfer capability", *International Transactions on Electrical Energy Systems*, 2021, 31(11), e13031, pp. 1–14, doi:10.1002/2050-7038.13031.

Eidiani, M., "A reliable and efficient method for assessing voltage stability in transmission and distribution networks", *International Journal of Electrical Power and Energy Systems*, 2011, 33(3), pp. 453–456, doi: 10.1016/j.ijepes.2010.10.007.

Eidiani, M., "An efficient differential equation load flow method to assess dynamic available transfer capability with wind farms", *IET Renewable Power Generation*, 2021, pp. 3843–3855, doi:10.1049/rpg2.12299.

Eidiani, M., "Assessment of voltage stability with new NRS," *2008 IEEE 2nd International Power and Energy Conference*, 2008, pp. 494–496, doi: 10.1109/PECON.2008.4762525.

Eidiani, M., "ATC evaluation by CTSA and POMP, two new methods for direct analysis of transient stability," *IEEE/PES Transmission and Distribution Conference and Exhibition*, 2002, pp. 1524–1529, vol. 3, doi: 10.1109/TDC.2002.1176824.

Eidiani, M., A. Ghavami, "New network design for simultaneous use of electric vehicles, photovoltaic generators, wind farms and energy storage," *2022 9th Iranian Conference on Renewable Energy & Distributed Generation (ICREDG)*, 2022, pp. 1–5, doi: 10.1109/ICREDG54199.2022.9804534.

Eidiani, M., A. Ghavami, H. Zeynal, Z. Zakaria, "Comparative analysis of mono-facial and bifacial photovoltaic modules for practical grid-connected solar power plant using PVsyst", *2022 IEEE International Conference on Power and Energy (PECon2022)*, Langkawi, Kedah, Malaysia, 6 December, 2022.

Eidiani, M., D. Yazdanpanah, "Minimum distance, a quick and simple method of determining the static ATC", *Journal of Electrical Engineering*, 2011, 11(2), pp. 95–101.

Eidiani, M., H. Zeynal, "A fast Holomorphic method to evaluate available transmission capacity with large scale wind turbines," *2022 9th Iranian Conference on Renewable Energy & Distributed Generation (ICREDG)*, 2022, pp. 1–5, doi: 10.1109/ICREDG54199.2022.9804527.

Eidiani, M., H. Zeynal, "Determination of online DATC with uncertainty and state estimation," *2022 9th Iranian Conference on Renewable Energy & Distributed Generation (ICREDG)*, 2022, pp. 1–6, doi: 10.1109/ICREDG54199.2022.9804581.

Eidiani, M., H. Zeynal, "New approach using structure-based modeling for the simulation of real power/frequency dynamics in deregulated power systems", *Turkish Journal of Electrical Engineering and Computer Sciences*, 2014, 22(5), pp. 1130–1146, doi:10.3906/elk-1208-90.

Eidiani, M., H. Zeynal, A.K. Zadeh, K.M. Nor, "Exact and efficient approach in static assessment of Available Transfer Capability (ATC)," *2010 IEEE International Conference on Power and Energy*, 2010, pp. 189–194, doi: 10.1109/PECON.2010.5697580.

Eidiani, M., H. Zeynal, A.K. Zadeh, S. Mansoorzadeh, K.M. Nor, "Voltage stability assessment: An approach with expanded Newton Raphson-Sydel," *2011 5th International Power Engineering and Optimization Conference*, 2011, pp. 31–35, doi: 10.1109/PEOCO.2011.5970424.

Eidiani, M., H. Zeynal, M. Shaaban, "A detailed study on prevailing ATC methods for optimal solution development", *2022 IEEE International Conference on Power and Energy (PECon2022)*, Langkawi, Kedah, Malaysia, 6 December. 2022.

Eidiani, M., H. Zeynal, Z. Zakaria, "An efficient Holomorphic based available transfer capability solution in presence of large scale wind farms," *2022 IEEE International Conference in Power Engineering Application (ICPEA)*, 2022, pp. 1–5, doi: 10.1109/ICPEA53519.2022.9744711.

Eidiani, M., H. Zeynal, Z. Zakaria, "Development of online dynamic ATC calculation integrating state estimation," *2022 IEEE International Conference in Power Engineering Application (ICPEA)*, 2022, pp. 1–5, doi: 10.1109/ICPEA53519.2022.9744694.

Eidiani, M., H. Zeynal, Z. Zakaria, M. Shaaban, "A comprehensive study on the renewable energy integration using DIgSILENT", *2023 IEEE 3rd International Conference in Power Engineering Applications (ICPEA)*, Putrajaya, Malaysia, 6–7 March 2023.

Eidiani, M., H. Zeynal, Z. Zakaria, M. Shaaban, "Analysis of optimization methods applied for renewable energy integration", *2023 IEEE 3rd International Conference in Power Engineering Applications (ICPEA)*, Putrajaya, Malaysia, 6–7 March 2023.

Eidiani, M., M. Ebrahimean Baydokhty, M. Ghamat, H. Zeynal, H. Mortazavi, "Transient stability enhancement via hybrid technical approach," *2011 IEEE Student Conference on Research and Development*, 2011, pp. 375–380, doi: 10.1109/SCOReD.2011.6148768.

Eidiani, M., M.H.M. Shanechi, E. Vaahedi, "Fast and accurate method for computing FCTTC (first contingency total transfer capability)," *Proceedings of International Conference on Power System Technology*, 2002, pp. 1213–1217, vol. 2, doi: 10.1109/ICPST.2002.1047595.

Eidiani, M., M. Kargar, "Frequency and voltage stability of the microgrid with the penetration of renewable sources," *2022 9th Iranian Conference on Renewable Energy & Distributed Generation (ICREDG)*, 2022, pp. 1–6, doi: 10.1109/ICREDG54199.2022.9804542.

Eidiani, M., M.E. Badokhty, M. Ghamat, H. Zeynal, "Improving transient stability using combined generator tripping and braking resistor approach", *International Review on Modelling and Simulations*, 2011, 4(4), pp. 1690–1699.

Eidiani, M., M.H.M. Shanechi, "FAD-ATC: A new method for computing dynamic ATC", *International Journal of Electrical Power and Energy Systems*, 2006, 28(2), pp. 109–118, doi:10.1016/j.ijepes.2005.11.004.

Eidiani, M., M.O. Buygi, S. Ahmadi, "CTV, complex transient and voltage stability: A new method for computing dynamic ATC", *International Journal of Power and Energy Systems*, 2006, 26(3), pp. 296–304, doi:10.2316/Journal.203.2006.3.203-3597.

Eidiani, M., N. Asghari Shahdehi, H. Zeynal, "Improving dynamic response of wind turbine driven DFIG with novel approach," *2011 IEEE Student Conference on Research and Development*, 2011, pp. 386–390, doi: 10.1109/SCOReD.2011.6148770.

Eidiani, M., S.M. Asadi, S.A. Faroji, M.H. Velayati, D. Yazdanpanah, "Minimum distance, a quick and simple method of determining the static ATC," *2008 IEEE 2nd International Power and Energy Conference*, 2008, pp. 490–493, doi: 10.1109/PECON.2008.4762524.

Eidiani, M., Y. Ashkhane, M. Khederzadeh, "Reactive power compensation in order to improve static voltage stability in a network with wind generation," *2009 International Conference on Computer and Electrical Engineering, ICCEE 2009*, 2009, pp. 47–50, doi: 10.1109/ICCEE.2009.239.

Ghardashi, G., M. Gandomkar, S. Majidi, M. Eidiani, S. Dadfar, "Accuracy and speed improvement of microgrid islanding detection based on PV using frequency-reactive power feedback method," *2022 International Conference on Protection and Automation of Power Systems (IPAPS)*, 2022, pp. 1–8, doi: 10.1109/IPAPS55380.2022.9763190.

Zeynal, H., A.K. Zadeh, K.M. Nor, M. Eidiani, "Locational Marginal Price (LMP) assessment using hybrid active and reactive cost minimization", *International Review of Electrical Engineering*, 2010, 5(5), pp. 2413–2418.

Zeynal, H., L.X. Hui, Y. Jiazhen, M. Eidiani, B. Azzopardi, "Improving Lagrangian relaxation unit commitment with Cuckoo search algorithm," *2014 IEEE International Conference on Power and Energy (PECon)*, 2014, pp. 77–82, doi: 10.1109/PECON.2014.7062417.

Zeynal, H., M. Eidiani, "Hydrothermal scheduling flexibility enhancement with pumped-storage units," *2014 22nd Iranian Conference on Electrical Engineering (ICEE)*, 2014, pp. 820–825, doi: 10.1109/IranianCEE.2014.6999649.

Zeynal, H., M. Eidiani, D. Yazdanpanah, "Intelligent substation automation systems for robust operation of smart grids," *2014 IEEE Innovative Smart Grid Technologies - Asia (ISGT ASIA)*, 2014, pp. 786–790, doi: 10.1109/ISGT-Asia.2014.6873893.

Zeynal, H., Y. Jiazhen, B. Azzopardi, M. Eidiani, "Flexible economic load dispatch integrating electric vehicles," *2014 IEEE 8th International Power Engineering and Optimization Conference (PEOCO2014)*, 2014, pp. 520–525, doi: 10.1109/PEOCO.2014.6814484.

Zeynal, H., Y. Jiazhen, B. Azzopardi, M. Eidiani, "Impact of electric vehicle's integration into the economic VAr dispatch algorithm," *2014 IEEE Innovative Smart Grid Technologies - Asia (ISGT ASIA)*, 2014, pp. 780–785, doi: 10.1109/ISGT-Asia.2014.6873892.

Index

Printed in the United States
by Baker & Taylor Publisher Services